tsixE toN oD srebmuN evitageN

Negative numbers do not exist

Martin Erik Horn

Martin Erik Horn
Lilienthalpark
D – 12209 Berlin
Germany
mail@martinerikhorn.de

nroH kirE nitraM
kraplahtneiliL
nilreB 12209 – D
ynamreG
ed.nrohkirenitram@liam

srebmuN evitageN
tsixE toN oD

Negative numbers do not exist

Martin Erik Horn

Bibliografische Information der Deutschen Nationalbibliothek:

Die Deutsche Nationalbibliothek verzeichnet diese Publikation
in der Deutschen Nationalbibliografie;
detaillierte bibliografische Daten sind im Internet über

http://dnb.dnb.de

im Internet abrufbar.

© 2022 Martin Erik Horn

Herstellung und Verlag:
BoD – Books on Demand, Norderstedt

ISBN: 978-3-7562-5832-1

stnetnoC

„noitcidartnoc lanretni na era srebmun evitageN“
“.timda snaicitamehtam lla sa

[1] regnihiaV snaH

dniknaM olleH   0

!sgnieb ralugnatcer olleH !snamuh olleH
,erutcurts lacigoloib ruoy otni detaroprocni era selgna thgiR
.yrtemoeg lanretni ruoy otni
.rennam ralugnatcer a ni gnitaluclac era uoy eroferehT
.snrettap detneiro tfel susrev thgir rof deniart era sniarb ruoy eroferehT
.srebmun evitagen detnevni evah uoy eroferehT

.elgna thgir ruoy si sihT

.noitcerid evitisop a otni thgir eht ot gnitniop si smra ruoy fo enO
,noitcerid evitagen a otni tfel eht ot gnitniop si mra dnoces ruoy dnA

.sdrawnwod ro sdrawpu smra ruoy ot ralucidneprep detneiro era sgel dna kcen elihw

.yrtemmys thgir tfel laretalib a evah ton od sehsifrats tnegilletni taht htiw derapmoC
.rennam laretalirt a ni derutcurts era sehsifrats tnegilletnI
.snoitcerid eerht otni tniop yllacirtemmys hcihw smra elcatnet eerht ssessop yehT

.selgna 120° tuoba syawla kniht sehsifrats tnegilletni eroferehT
.srebmun evitagen tnevni ton od sniarb riehT
.scitamehtam evitisop dloferht a no desab gnikrow era sniarb riehT
.tsixe ton od srebmun evitageN
.snoitcerid evitisop eerht eht fo hcae rof srebmun evitisop ylno era erehT

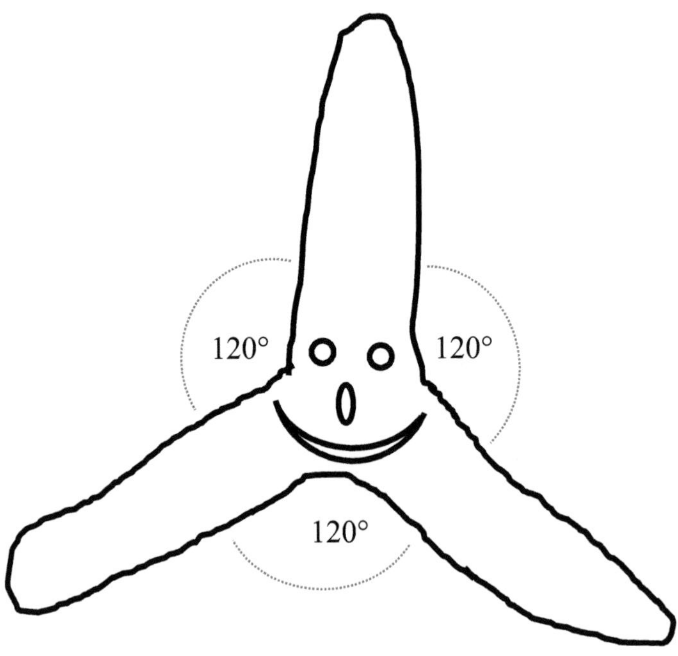

srebmuN evitisoP 1

.tsixe ton od srebmun evitageN
.emit fo htgnel evitagen a derusaem reve sah ,gnieb namuh on ,ydoboN
.ecaps fo ecnatsid evitagen a derusaem reve sah ,gnieb namuh on ,ydobon dnA
relur eht dnuora denrut ylno evah niaga dna niaga su fo llA
.noitcerid rehtona otni secnatsid evitisop derusaem neht dna
.evitisop si ,gnirusaem era ew tahw ,gnihtyrevE

.tsixe ton od srebmun evitageN
.tsixe od srebmun evitisoP
.eno rebmun eht fo selpitlum era srebmun evitisoP
:eno rebmun eht si siht ,ereh dnA

$$1 = \begin{pmatrix} 1 & 0 & 0 \\ 0 & 1 & 0 \\ 0 & 0 & 1 \end{pmatrix}$$

.enin ot lauqe si xis sulp eerhT

$$3 + 6 = \begin{pmatrix} 3 & 0 & 0 \\ 0 & 3 & 0 \\ 0 & 0 & 3 \end{pmatrix} + \begin{pmatrix} 6 & 0 & 0 \\ 0 & 6 & 0 \\ 0 & 0 & 6 \end{pmatrix} = \begin{pmatrix} 9 & 0 & 0 \\ 0 & 9 & 0 \\ 0 & 0 & 9 \end{pmatrix} = 9$$

.thgie ot lauqe si owt semit ruof dnA

$$4 \cdot 2 = \begin{pmatrix} 4 & 0 & 0 \\ 0 & 4 & 0 \\ 0 & 0 & 4 \end{pmatrix} \begin{pmatrix} 2 & 0 & 0 \\ 0 & 2 & 0 \\ 0 & 0 & 2 \end{pmatrix} = \begin{pmatrix} 8 & 0 & 0 \\ 0 & 8 & 0 \\ 0 & 0 & 8 \end{pmatrix} = 8$$

:noisulcnoC

!arbegla xirtam nrael s'teL
.srebmun evitagen deen ton od eW
.yrassecennu era yehT
.suoulfrepus era yehT
.tsixe ton od yehT

seititnE cisaB   2

.tsixe seititne lacitamehtam cisab xis rehtegotlA

:era yeht ereH

:eno rebmun ehT
$$1 = \begin{pmatrix} 1 & 0 & 0 \\ 0 & 1 & 0 \\ 0 & 0 & 1 \end{pmatrix}$$

:thgir eht ot gnitniop ,rotcev tsriF
$$e_1 = \begin{pmatrix} 1 & 0 & 0 \\ 0 & 0 & 1 \\ 0 & 1 & 0 \end{pmatrix}$$

:tfel eht ot sdrawpu gnitniop ,rotcev dnoceS
$$e_2 = \begin{pmatrix} 0 & 0 & 1 \\ 0 & 1 & 0 \\ 1 & 0 & 0 \end{pmatrix}$$

:tfel eht ot sdrawnwod gnitniop ,rotcev drihT
$$e_3 = \begin{pmatrix} 0 & 1 & 0 \\ 1 & 0 & 0 \\ 0 & 0 & 1 \end{pmatrix}$$

.120° si syawla srotcev eerht eseht neewteb elgna ehT

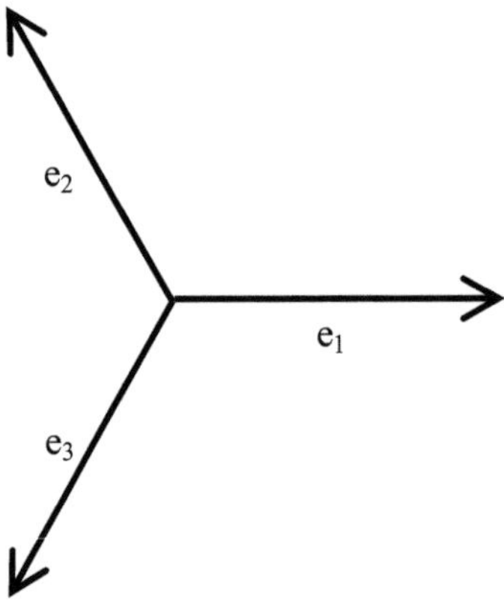

.rats sedecrem a ekil skool wohemos sihT
.dopirt enalp a ,rats sedecrem euqilbo na si ti tuB
.retal emoc neht lliw sdoparteT

,eno ot erauqs yeht sA
.seititne tinu era seititne cisab ruof eseht

$$e_1{}^2 = e_2{}^2 = e_3{}^2 = 1^2 = \begin{pmatrix} 1 & 0 & 0 \\ 0 & 1 & 0 \\ 0 & 0 & 1 \end{pmatrix} = 1$$

,seititne cisab lanigiro eht ot lacitnedi era sesrevni rieht sa dnA
seititne cisab eseht yb edivid ot elba era ew

$$1^{-1} = \begin{pmatrix} 1 & 0 & 0 \\ 0 & 1 & 0 \\ 0 & 0 & 1 \end{pmatrix}$$

$$e_1{}^{-1} = \begin{pmatrix} 1 & 0 & 0 \\ 0 & 0 & 1 \\ 0 & 1 & 0 \end{pmatrix} \qquad e_2{}^{-1} = \begin{pmatrix} 0 & 0 & 1 \\ 0 & 1 & 0 \\ 1 & 0 & 0 \end{pmatrix} \qquad e_3{}^{-1} = \begin{pmatrix} 0 & 1 & 0 \\ 1 & 0 & 0 \\ 0 & 0 & 1 \end{pmatrix}$$

esuaceb

$$e_1\, e_1{}^{-1} = e_2\, e_2{}^{-1} = e_3\, e_3{}^{-1} = 1 \cdot 1^{-1} = \begin{pmatrix} 1 & 0 & 0 \\ 0 & 1 & 0 \\ 0 & 0 & 1 \end{pmatrix} = 1$$

eno ni stluser ylpmis siht lla
.noitacilpitlum fo tnemele ytitnedi ro tnemele lartuen eht si hcihw

!arbegla xirtam evil gnoL

oreZ 3

.tsixe ton od srebmun evitageN
(orez fo noitisop eht ot suht dna) nigiro eht ot kcab og oT
,thgir eht ot e_1 fo noitcerid evitisop eht otni pets elgnis a edam gnivah retfa
.tfel eht ot pets-eno-sunim a mrofrep ot elba ton won era ew
.tsixe ton seod ylpmis pets-eno-sunim evitagen A

,niaga orez hcaer oT
e_2 fo noitcerid evitisop eht otni pets eno og ot evah daetsni ew
.e_3 fo noitcerid eht otni pets evitisop rehtona dna

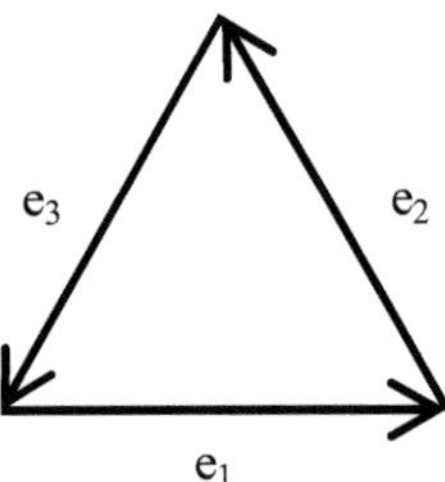

:orez ni stluser yllanif e_3 sulp e_2 sulp e_1

$$e_1 + e_2 + e_3 = 0$$

:sdaer noitauqe siht secirtam gnisU

$$e_1 + e_2 + e_3 = \begin{pmatrix} 1 & 0 & 0 \\ 0 & 0 & 1 \\ 0 & 1 & 0 \end{pmatrix} + \begin{pmatrix} 0 & 0 & 1 \\ 0 & 1 & 0 \\ 1 & 0 & 0 \end{pmatrix} + \begin{pmatrix} 0 & 1 & 0 \\ 1 & 0 & 0 \\ 0 & 0 & 1 \end{pmatrix} = \begin{pmatrix} 1 & 1 & 1 \\ 1 & 1 & 1 \\ 1 & 1 & 1 \end{pmatrix} = 0$$

.aes cariD eht fo ruoivaheb eht ot ralimis si sihT
.ereh eb lliw gnihton ,deipucco era snoitisop lla fI
.orez fo etats eht ot lacitnedi si sihT
.eno ot lacitnedi si won xirtam eht fo tnemele yrevE

.yaw evitanretla na ni orez etirw ot elbisop osla si ti esruoc fO
pets on spahrep ro spets eerht ro spets owt og thgim ew ,elpmaxe roF
.snoitcerid eerht eht fo hcae otni

:secirtam orez tnereffid ynam yletinifni yb detneserper eb nac orez rebmun eht suhT

$$\begin{pmatrix}1&1&1\\1&1&1\\1&1&1\end{pmatrix}=\begin{pmatrix}2&2&2\\2&2&2\\2&2&2\end{pmatrix}=\begin{pmatrix}3&3&3\\3&3&3\\3&3&3\end{pmatrix}=\ldots=\begin{pmatrix}0&0&0\\0&0&0\\0&0&0\end{pmatrix}=0$$

secirtam orez tnereffid eht mrofsnart ot esnes sekam syawla ti ytiralc fo ekas eht roF
.(thgir eht no ees ,stnemele orez htiw) xirtam orez dradnats eht otni
snoitaluclac lanoitevnoc gnimrofrep nehw yaw ralimis a ni siht gniod era eW
.orez fo selpitlum lla tcelgen ylpmis ew :scitamehtam lanoitnevnoc ,lausu ni

$$7\,e_1+5\,e_2+6\,e_3=\begin{pmatrix}7&0&0\\0&0&7\\0&7&0\end{pmatrix}+\begin{pmatrix}0&0&5\\0&5&0\\5&0&0\end{pmatrix}+\begin{pmatrix}0&6&0\\6&0&0\\0&0&6\end{pmatrix}$$

$$=\begin{pmatrix}7&6&5\\6&5&7\\5&7&6\end{pmatrix}=\underbrace{5\,e_1+5\,e_2+5\,e_3}_{0}+2\,e_1+e_3=5\cdot0+2\,e_1+e_3$$

$$=\begin{pmatrix}2&1&0\\1&0&2\\0&2&1\end{pmatrix}=2\,e_1+e_3$$

,derongi eb nac orez semit evif sA
:thgir eht ot sdrawnwod stniop hcihw ,rotcev a eb lliw tluser eht

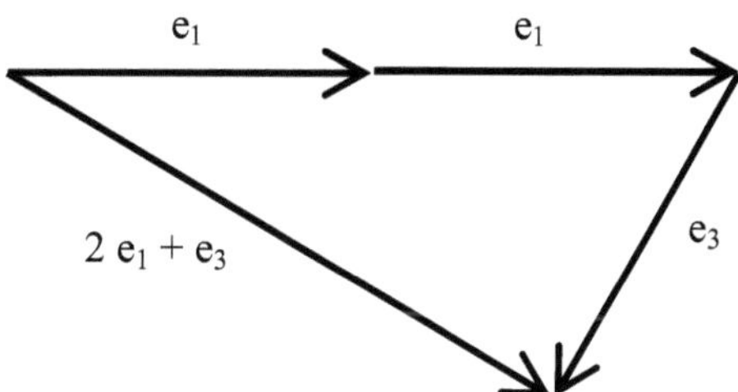

:gnirauqs yb dnuof eb neht nac rotcev siht fo htgnel ehT

$$(2\,e_1+e_3)^2=\begin{pmatrix}2&1&0\\1&0&2\\0&2&1\end{pmatrix}^2=\begin{pmatrix}5&2&2\\2&5&2\\2&2&5\end{pmatrix}=\begin{pmatrix}3&0&0\\0&3&0\\0&0&3\end{pmatrix}=3$$

Here again $0 \cdot 2 = \begin{pmatrix} 2 & 2 & 2 \\ 2 & 2 & 2 \\ 2 & 2 & 2 \end{pmatrix}$ has been ignored to get the standard representation.

The square root of this result is then identical to the length of the vector:

$$|2\,e_1 + e_3| = \sqrt{(2\,e_1 + e_3)^2} = \sqrt{3} \approx 1.73$$

Checking this result by applying the Theorem of Pythagoras

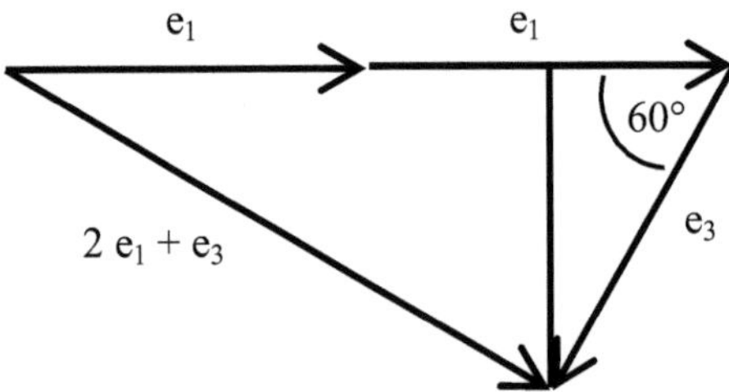

will give the identical value:

$$(2\,e_1 + e_3)^2 = (2 - \cos 60°)^2 + (\sin 60°)^2 = 2.25 + 0.75 = 3$$

$\Leftarrow$ The length of the vector $(2\,e_1 + e_3)$ is thus $\sqrt{3}$ units of length.

Thus the matrix term

$$\begin{pmatrix} 2 & 2 & 2 \\ 2 & 2 & 2 \\ 2 & 2 & 2 \end{pmatrix} = 2 \begin{pmatrix} 0 & 0 & 0 \\ 0 & 0 & 0 \\ 0 & 0 & 0 \end{pmatrix} = 2 \cdot 0$$

was indeed neglected successfully without unexpected complications.

4 Products of Two Vectors

The last two basic entities can be constructed
by multiplying two different unit vectors.

$$\text{First rhombus:} \qquad e_{12} = e_1 e_2 = e_2 e_3 = e_3 e_1 = \begin{pmatrix} 0 & 0 & 1 \\ 1 & 0 & 0 \\ 0 & 1 & 0 \end{pmatrix}$$

$$\text{Second rhombus:} \qquad e_{21} = e_2 e_1 = e_3 e_2 = e_1 e_3 = \begin{pmatrix} 0 & 1 & 0 \\ 0 & 0 & 1 \\ 1 & 0 & 0 \end{pmatrix}$$

These multiplications of different unit vectors
result in rhombuses with interior angles of 60° or 120° respectively.

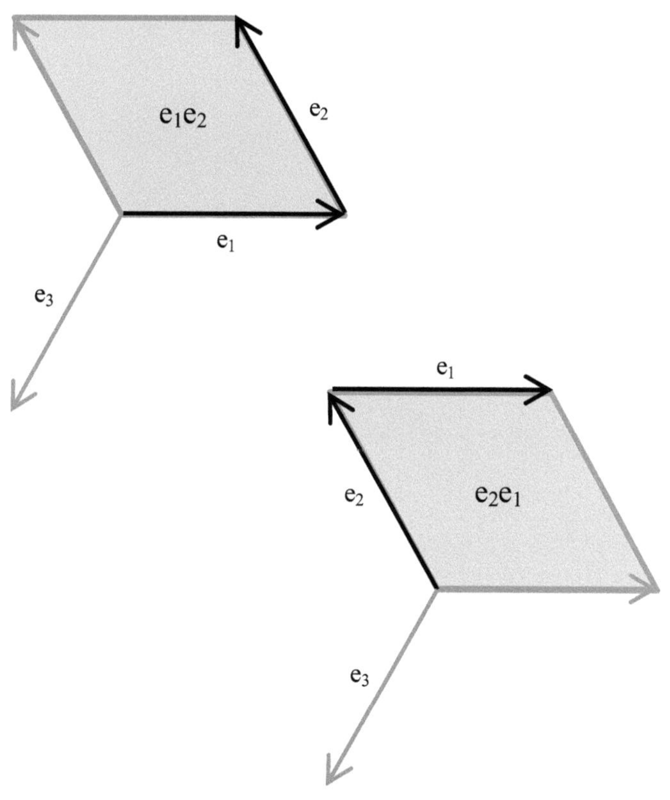

From the viewpoint of Geometric Algebra [2], [3], [4] the three rhombuses

$$e_{12} = e_1 e_2 \qquad e_{23} = e_2 e_3 \qquad e_{31} = e_3 e_1$$

are completely identical, because they possess an identical area (with identical lengths of all sides), identical interior angles and identical orientation.

And the three rhombuses

$$e_{21} = e_2 e_1 \qquad e_{32} = e_3 e_2 \qquad e_{13} = e_1 e_3$$

are completely identical as well.

Although being positioned differently, their area properties are absolutely identical

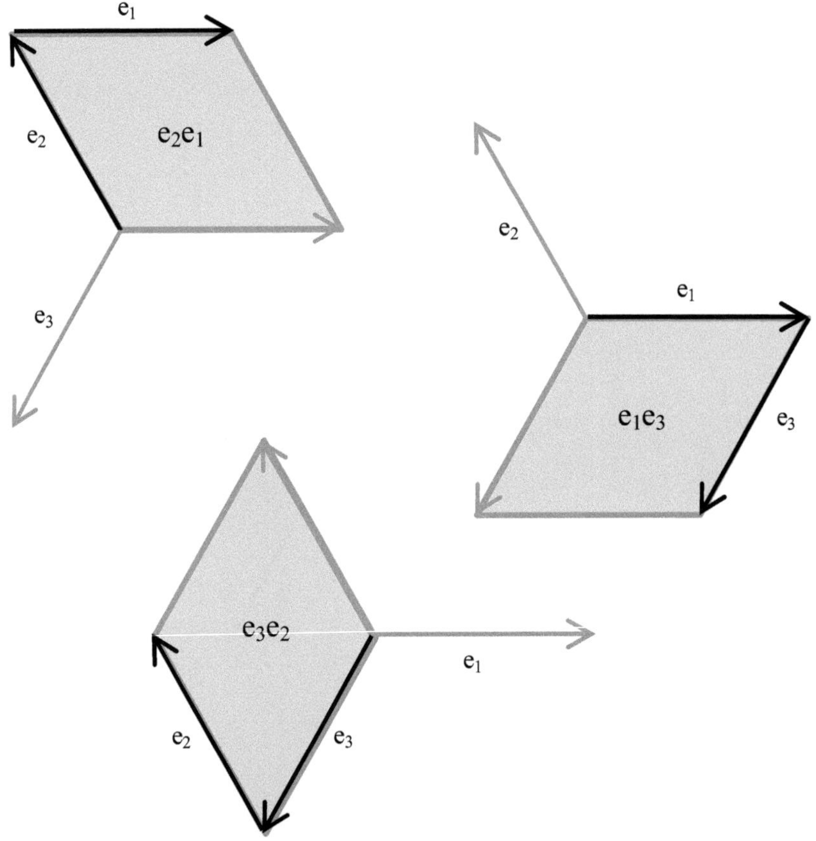

Between these rhombuses no differences exist with respect to the area.
Mathematically they are absolutely identical!

These last two basic entities square in a cyclical exchanged order.
The first anti-clockwise oriented rhombus e_{12}

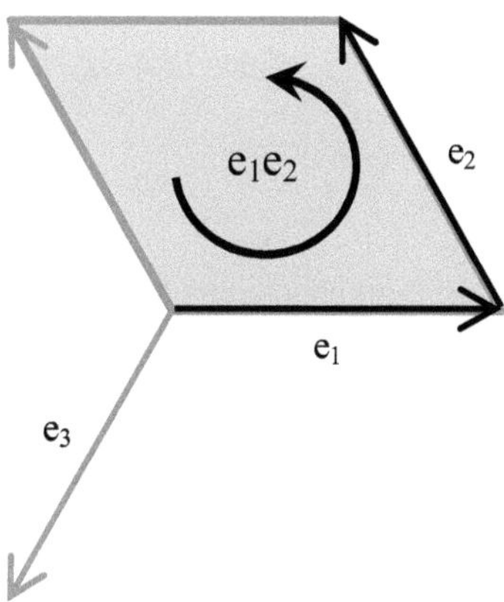

will be transformed

$$e_{12}{}^2 = \begin{pmatrix} 0 & 0 & 1 \\ 1 & 0 & 0 \\ 0 & 1 & 0 \end{pmatrix}^2 = \begin{pmatrix} 0 & 1 & 0 \\ 0 & 0 & 1 \\ 1 & 0 & 0 \end{pmatrix} = e_{21}$$

into the second clockwise oriented rhombus e_{21} by squaring.

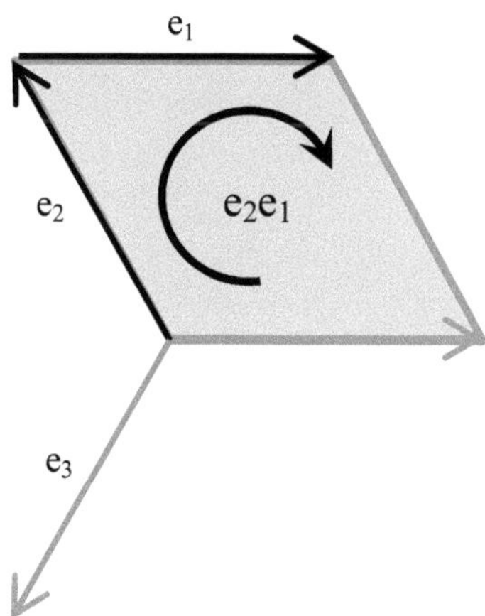

demrofsnart eb lliw e_{21} submohr detneiro esiwkcolc dnoces eht dnA

$$e_{21}{}^2 = \begin{pmatrix} 0 & 1 & 0 \\ 0 & 0 & 1 \\ 1 & 0 & 0 \end{pmatrix}^2 = \begin{pmatrix} 0 & 0 & 1 \\ 1 & 0 & 0 \\ 0 & 1 & 0 \end{pmatrix} = e_{12}$$

.gnirauqs yb e_{12} submohr detneiro esiwkcolc-itna tsrif eht otni

mus llun eht ,yaw eht yb ,sey hO

$$e_1 + e_2 + e_3 = 0$$

.(e_3 yb ro e_2 yb ro) e_1 yb deilpitlum eb osla nac
.retpahc txen eht ni enod eb lliw siht dnA

evitisoP si enO evitageN 5

:3 retpahc ni desylana neeb ydaerla sah 0 rotcev orez ehT

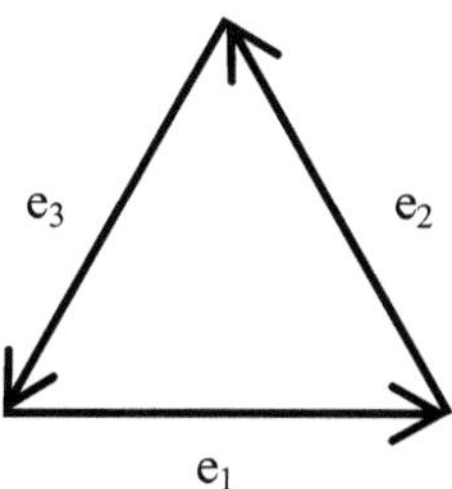

:orez eb tsum tluser eht ,dedda era e_3 dna e_2 dna e_1 fI

$$e_1 + e_2 + e_3 = 0$$

noitauqe eht ,snaicitamehtam ereves dna tcirts fo dlrow lacigol eht nI

$$e_1 + (- e_1) = 0$$

:swollof ti suhT .sdloh osla

$$(- e_1) = e_2 + e_3$$

:ylsuogolana yletelpmoc dnA

$$(- e_2) = e_3 + e_1$$

$$(- e_3) = e_1 + e_2$$

,sraeppa rotcev evitagen a erehwemos revenehW
yltnatsni dna noitatiseh tuohtiw yletaidemmi ti ecalper lliw ew
.srotcev owt rehto eht fo noitanibmoc raenil evitisop a yb

,detsixe dah yeht fi dnA) .tsixe ton od srotcev evitageN
(!yaw evitisop a ni yltnatsni dna yletaidemmi meht nettirwer evah dluow ew

.srebmun evitagen htiw nekatrednu eb won lliw erudecorp emas ehT
.e_1 yb $e_1 + e_2 + e_3 = 0$ mus llun eht ylpitlum ew taht roF

$$e_1 (e_1 + e_2 + e_3) = e_1\, 0$$

$$e_1{}^2 + e_1e_2 + e_1e_3 = e_1 \ 0 \qquad\qquad \Leftarrow$$

$$1 + e_{12} + e_{21} = 0 \qquad\qquad \Leftarrow$$

derapmoc eb won nac tluser sihT
.sngis evitagen gnivol snaicitamehtam fo dlrow lacigol ereves dna tcirts eht htiw

$$1 + (-1) \ = \ 0$$

:dedulcnoc eb nac ti dnA

$$(-1) = e_{12} + e_{21}$$

,sraeppa rebmun evitagen a erehwemos revenehW
yltnatsni dna noitatiseh tuohtiw yletaidemmi ti ecalper lliw ew
.sesubmohr owt eht fo noitanibmoc raenil evitisop a yb

.tsixe ton od srebmun evitageN
,detsixe dah yeht fi dnA
yltnatsni dna yletaidemmi meht decalper evah dluow ew
:erutcurts gniwollof eht fo stnemele evitisop ylno htiw secirtam 3 x 3 yb

$$(-1) = e_{12} + e_{21} = \begin{pmatrix} 0 & 1 & 1 \\ 1 & 0 & 1 \\ 1 & 1 & 0 \end{pmatrix}$$

.aes cariD eht fo ruoivaheb eht ot ralimis si sihT
.aes cariD eht fo etats orez deipucco yletelpmoc eht ni seloh era selcitrap-itnA

-yreve seno htiw $\begin{pmatrix} 1 & 1 & 1 \\ 1 & 1 & 1 \\ 1 & 1 & 1 \end{pmatrix} = \begin{pmatrix} 0 & 0 & 0 \\ 0 & 0 & 0 \\ 0 & 0 & 0 \end{pmatrix} = 0$ xirtam llun ro xirtam orez eht tA

snoitisop eht ta seloh sah won $\begin{pmatrix} 0 & 1 & 1 \\ 1 & 0 & 1 \\ 1 & 1 & 0 \end{pmatrix} = e_{21} + e_{12}$ eno-itna eht ,erehw

$\begin{pmatrix} 1 & 0 & 0 \\ 0 & 1 & 0 \\ 0 & 0 & 1 \end{pmatrix} = 1$ xirtam ytitnedi ro xirtam tinu eht fo seno eht fo

.ytitne evitisop a si ylerus eno-itna siht tniopweiv lacihposolihp a morF
.snoitatneiro desoppo htiw e_{21} dna e_{12} sesubmohr owt eht fo mus eht si tI

!evitisop si (−1) eno-itna ehT

6 Multivectors

Thus we are now able to express every vector

$$\mathbf{r} = x_1\,e_1 + x_2\,e_2 + x_3\,e_3 = \begin{pmatrix} x_1 & x_3 & x_2 \\ x_3 & x_2 & x_1 \\ x_2 & x_1 & x_3 \end{pmatrix}$$

as a linear combination of only two of the three unit vectors e_1, e_2, e_3.
Then two of the three components x_1, x_2, x_3 are positive
and one of the three components is zero.

$$\mathbf{r} = x_1\,e_1 + x_2\,e_2 = \begin{pmatrix} x_1 & 0 & x_2 \\ 0 & x_2 & x_1 \\ x_2 & x_1 & 0 \end{pmatrix} \qquad \Leftarrow \qquad x_1 > 0; \quad x_2 > 0; \quad x_3 = 0$$

$$\mathbf{r} = x_1\,e_1 + x_3\,e_3 = \begin{pmatrix} x_1 & x_3 & 0 \\ x_3 & 0 & x_1 \\ 0 & x_1 & x_3 \end{pmatrix} \qquad \Leftarrow \qquad x_1 > 0; \quad x_2 = 0; \quad x_3 > 0 \qquad \text{or}$$

$$\mathbf{r} = x_2\,e_2 + x_3\,e_3 = \begin{pmatrix} 0 & x_3 & x_2 \\ x_3 & x_2 & 0 \\ x_2 & 0 & x_3 \end{pmatrix} \qquad \Leftarrow \qquad x_1 = 0; \quad x_2 > 0; \quad x_3 > 0 \qquad \text{or}$$

– if the vector is not lying parallel to one of the three axes.
In this case two components will be zero.

In a similar way the linear combination of scalar and rhombuses

$$\mathbf{R} = x_0\,1 + x_{12}\,e_{12} + x_{21}\,e_{21} = \begin{pmatrix} x_0 & x_{21} & x_{12} \\ x_{12} & x_0 & x_{21} \\ x_{21} & x_{12} & x_0 \end{pmatrix}$$

can be reduced to only two positive components

$$\mathbf{R} = x_0 + x_{12}\,e_{12} = \begin{pmatrix} x_0 & 0 & x_{12} \\ x_{12} & x_0 & 0 \\ 0 & x_{12} & x_0 \end{pmatrix} \qquad \Leftarrow \qquad x_0 > 0; \quad x_{12} > 0; \quad x_{21} = 0$$

$$\mathbf{R} = x_0 + x_{21}\,e_{21} = \begin{pmatrix} x_0 & x_{21} & 0 \\ 0 & x_0 & x_{21} \\ x_{21} & 0 & x_0 \end{pmatrix} \qquad \Leftarrow \qquad x_0 > 0; \quad x_{12} = 0; \quad x_{21} > 0 \qquad \text{or}$$

$$R = x_{12}\, e_{12} + x_{21}\, e_{21} = \begin{pmatrix} 0 & x_{21} & x_{12} \\ x_{12} & 0 & x_{21} \\ x_{21} & x_{12} & 0 \end{pmatrix} \quad \Leftarrow \quad x_0 = 0; \quad x_{12} > 0; \quad x_{21} > 0 \qquad \text{ro}$$

arbegla hsifrats evitisop ylerup siht nI
.tcejbo lacitamehtam detacilpmoc tsom eht eb lliw **M** rotcevitlum eht

$$M = r + R = x_0 + x_1\, e_1 + x_2\, e_2 + x_3\, e_3 + x_{12}\, e_{12} + x_{21}\, e_{21}$$

$$= \begin{pmatrix} x_0 + x_1 & x_3 + x_{21} & x_2 + x_{12} \\ x_3 + x_{12} & x_0 + x_2 & x_1 + x_{21} \\ x_2 + x_{21} & x_1 + x_{12} & x_0 + x_3 \end{pmatrix}$$

sehsifrats tnegilletni fo dlrow lacitamehtam lanoisnemid-owt eht suhT
.sesubmohr owt dna ,srotcev eerht ,ralacs a fo desopmoc si

orez eb lliw x_{21}, x_{12}, x_3, x_2, x_1, x_0 stnenopmoc xis eht fo owt esruoc fO
.noitatneserper dradnats eht ni evitisop eb lliw stnenopmoc ruof gniniamer eht dna

sesarhP 7

snaicitamehtam yb arbegla hsifrats dellac ton si arbegla siht yllatnedicnI
.arbegla noitatumrep S_3 tub

.secirtam 3 x 3 eht fo ruoivaheb lacitamehtam eht si siht rof nosaer ehT

(snoitatumrep ro) segnahcretni elbissop lla tneserper secirtam esehT
.C dna ,B ,A stcejbo tnereffid eerht fo

$$\begin{pmatrix} 1 & 0 & 0 \\ 0 & 0 & 1 \\ 0 & 1 & 0 \end{pmatrix} \begin{pmatrix} A \\ B \\ C \end{pmatrix} = \begin{pmatrix} A \\ C \\ B \end{pmatrix}$$:stcejbo owt tsal eht segnahcretni e_1

$$\begin{pmatrix} 0 & 0 & 1 \\ 0 & 1 & 0 \\ 1 & 0 & 0 \end{pmatrix} \begin{pmatrix} A \\ B \\ C \end{pmatrix} = \begin{pmatrix} C \\ B \\ A \end{pmatrix}$$:tcejbo tsal eht dna tsrif eht segnahcretni e_2

$$\begin{pmatrix} 0 & 1 & 0 \\ 1 & 0 & 0 \\ 0 & 0 & 1 \end{pmatrix} \begin{pmatrix} A \\ B \\ C \end{pmatrix} = \begin{pmatrix} B \\ A \\ C \end{pmatrix}$$:stcejbo owt tsrif eht segnahcretni e_3

$$\begin{pmatrix} 0 & 0 & 1 \\ 1 & 0 & 0 \\ 0 & 1 & 0 \end{pmatrix} \begin{pmatrix} A \\ B \\ C \end{pmatrix} = \begin{pmatrix} C \\ A \\ B \end{pmatrix}$$:stcejbo eerht lla segnahcretni e_{12}

$$\begin{pmatrix} 0 & 1 & 0 \\ 0 & 0 & 1 \\ 1 & 0 & 0 \end{pmatrix} \begin{pmatrix} A \\ B \\ C \end{pmatrix} = \begin{pmatrix} B \\ C \\ A \end{pmatrix}$$:stcejbo eerht lla segnahcretni niaga e_{21}
(redro tnereffid ni tub)

$$\begin{pmatrix} 1 & 0 & 0 \\ 0 & 1 & 0 \\ 0 & 0 & 1 \end{pmatrix} \begin{pmatrix} A \\ B \\ C \end{pmatrix} = \begin{pmatrix} A \\ B \\ C \end{pmatrix}$$:gnihtyna egnahcretni ton seod 1 xirtam tinu eht dnA

.ffuts denoihsaf-dlo si siht lla tuB
srotcev nmuloc tceffe secirtam ereH
.detauqitna neve ro dedrater rehtar si siht dna –
secirtam dna srotcev nmuloc fo mrof ni seititnauq elbitapmocni owT
.rehtegot dehsalps era

!!kcmas, hsolps, hsalps, moob, gnaB

!ylgu woH !tnageleni woH !elbirroH
!suoived ylsuodnemert dna tneiciffeni woh : yllaicepse dnA

siht lla enod sah nnamssarG nnamreH
dednuof retteb yllacigol dna tnagele erom hcum ,lufituaeb erom hcum
.[6] ,[5] snoisnetxe fo yroeht sih ni

.erutcurts lacitnedi syawla fo stcejbo lacitamehtam sesu yltnatsisnoc nnamssarG
.erutcurts emas eht fo stcejbo lacitamehtam tceffe yeht dnA

secirtam dna srotcev fo erutxim gnisufnoc dna drusba on yroeht s'nnamssarG nI
,pu snrut
.detneserp si scitamehtam no weiv nredom yranoitulover dna tnetsisnoc a tub

,1844 ,snoisnetxe fo yroeht eht fo raeyhtrib ehT
scitamehtam fo raey rednow nedlog a ,[7] silibariM sunnA na neeb deedni sah
."dnim namuh eht fo yrotsih eht ni skramdnal emerpus eht fo eno" –

nnamssarG fo scitamehtam suollevram ,lufrednow siht gnisu yb dnA
egayov eht eunitnoc won lliw ew
.snoitaluclac fo dlrow gnisirprus ylsoulucarim eht hguorht
:wonk dluohs uoy roF
.srotarepo sa deredisnoc eb osla nac e_3 ,e_2, e_1 srotcev tinu eht

.stcudorp hciwdnas fo pleh eht htiw etarepo lliw ew retpahc gniwollof eht ni dnA

stcudorP hciwdnaS 8

:sdnarepo htiw stcejbo lacitamehtam ruo deifitnedi evah ylno ew won litnU
,no detca eb nac ti seititne eseht lla nO .sesubmohr detneiro ,srotcev ,sralacs

.seititne eseht egnahc ot dna tceffe ot snoitarepo fo mia eht si tI
:evitcepsrep tnereffid a morf seititne eseht otno gnikool won era ew eroferehT
?noitarepo na ta no detca nehw ytitne na fo gnignahc eht sesuac tahW

sa – sesubmohr detneiro ,srotcev ,sralacs – seititne emas eht esu won lliw eW
,adnarepo no gnitca era srotarepo fI .srotarepo
.degnahc eb lliw dna detceffe eb lliw sdnarepo eht

si ereht [14] ,[13] ,[12] ,[11] ,[10] ,[9] ,[8] arbeglA cirtemoeG nI
.tcudorp hciwdnas ehT :snoitarepo tuo gniyrrac rof nrettap elpmis yrev a

.elddim eht ni decalp si (degnahc eb lliw hcihw ytitne eht) dnarepo ehT
(noitarepo eht sesuac hcihw ytitne eht) rotarepo eht nehT
.thgir eht morf deilpitlum-tsop dna tfel eht morf deilpitlum-erp eb lliw
:ekil kool neht lliw sihT

dnarepo degnahc = rotarepo dnarepo rotarepo

,yaw tnacifingam dna lufituaeb yllaer a ni skrow ereh arbeglA cirtemoeG
.ytuaeb larutcurts gnitanicsaf dna gnicnivnoc a gniwohs
gnisirprus dna tnereffid eerht htiw sdnarepo sa tca seititne tnereffid eerht ehT
:gniwollof eht ni nevig era snoitarepo tnatropmi yrev esehT .stceffe

ezis fo egnahC .1

dnarepo reggib/rellams = ralacs dnarepo ralacs

noitcelfeR .2

dnarepo detcelfer = rotcev tinu dnarepo rotcev tinu

noitatoR .3

dnarepo detator = submohr tinu dnarepo submohr tinu

diputs emos .g.e) stcejbo lacitamehtam wen ynnuf ddo yna deen ton od ew suhT
.(secirtam noitcelfer diputs yllauqe emos ro secirtam noitator
.srotarepo sa seititnauq nwonk-llew ydaerla ruo esu ylpmis eW

!tcudorp hciwdnas eht evil gnoL

.tcelfer ot gniog era ew woN
:yaw elpmis yllaer a ni trats lliw eW
.detcelfer eb lliw rotcev tinu A

sixa na ni detcelfer si e_3 rotcev tinu ehT
.e_1 rotcev tinu fo noitcerid eht otni stniop hcihw

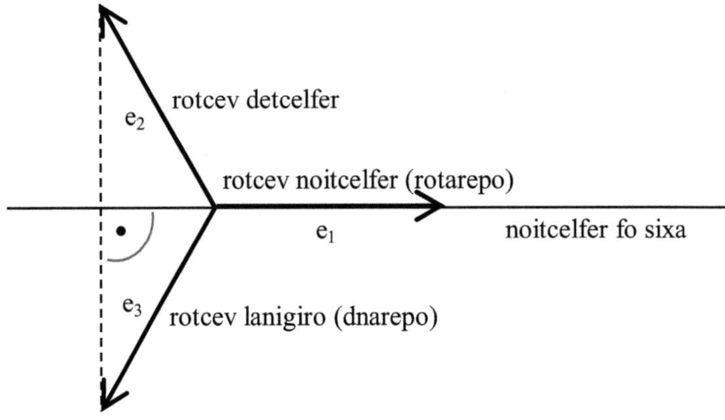

:si won tcudorp hciwdnas ehT

(rotcev noitcelfer) (rotcev lanigiro) (rotcev noitcelfer) = rotcev detcelfer

$$e_1\, e_3\, e_1 = e_1\, e_{12} = e_1\, e_1\, e_2 = e_1^{2}\, e_2 = e_2$$

sixa-e_1 eht ni detcelfer nehW
.e_2 rotcev tinu eht otni demrofsnart si e_3 rotcev tinu eht
.e_3 fo egami rorrim eht si e_2

sixa na ni detcelfer si e_3 rotcev tinu woN
.e_2 rotcev tinu fo noitcerid eht otni stniop hcihw

.elpmis etiuq si niaga sihT

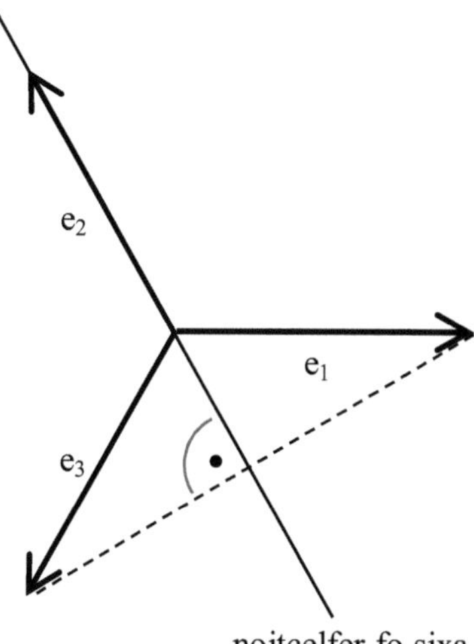

noitcelfer fo sixa

:si neht tcudorp hciwdnas ehT

(rotcev noitcelfer) (rotcev lanigiro) (rotcev noitcelfer) = rotcev detcelfer

$$e_2\ e_3\ e_2 = e_2\ e_{21} = e_2\ e_2\ e_1 = e_2{}^2\ e_1 = e_1$$

sixa-e_2 eht ni detcelfer nehW
.e_1 rotcev tinu eht otni demrofsnart si e_3 rotcev tinu eht
.e_3 fo egami rorrim eht si e_1 woN

:deredisnoc si esac driht eht yllaniF

sixa na ni detcelfer si e_3 rotcev tinu eht woN
.e_3 rotcev tinu emas eht fo noitcerid eht otni stniop hcihw

.neppah lliw gnihton ,tsixe ton seod tnenopmoc lanogohtro na won sA
.degnahcnu sniamer dna flesti otni detcelfer si e_3 rotcev tinu ehT

:tcudorp hciwdnas eht fo tluser sa nees eb osla nac sihT

$$e_3\ e_3\ e_3 = e_3{}^2\ e_3 = e_3$$

.egap gniwollof eht fo erugif eht ni nwohs si noitcelfer sihT

In addition this figure also shows that the other two unit vectors e_1 and e_2 are reflected reciprocally at the opposite unit vector.

These two sandwich products are then:

$$e_3\, e_1\, e_3 = e_3\, e_{21} = e_3\, e_3\, e_2 = e_3^{\,2}\, e_2 = e_2$$

$$e_3\, e_2\, e_3 = e_3\, e_{12} = e_3\, e_3\, e_1 = e_3^{\,2}\, e_1 = e_1$$

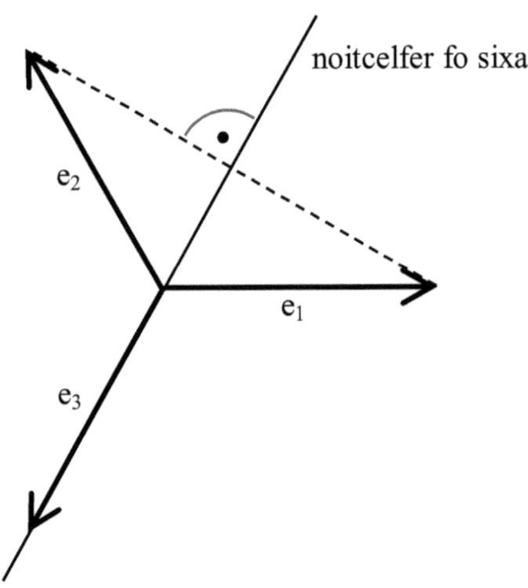

That's it.
Reflections really are very simple.
And they are linear with respect to the components.

For instance the reflection of the friendly unit vector $\mathbf{f}$

$$\mathbf{f} = \frac{5}{7}\, e_1 + \frac{8}{7}\, e_3 \approx 0.71\, e_1 + 1.14\, e_3 \qquad\qquad (\text{Yes, this is a unit vector!})$$

at an axis which points into the direction of unit vector e_2 can be split linearly into

$$\mathbf{f_{ref}} = e_2\, \mathbf{f}\, e_2 = e_2 \left(\frac{5}{7}\, e_1 + \frac{8}{7}\, e_3 \right) e_2 = \frac{5}{7}\, e_2\, e_1\, e_2 + \frac{8}{7}\, e_2\, e_3\, e_2 = \frac{8}{7}\, e_1 + \frac{5}{7}\, e_3$$

Thus different components of a vector can be reflected separately.

srotceV yrartibrA fo snoitcelfeR   10

uoy noitcerid yna otni stniop hcihw sixa na ni rotcev yrartibra na fo noitcelfer ehT
:mrof lareneg ni stcudorp hciwdnas tuoba retpahc eht ni denoitnem ydaerla saw ekil

(rotcev noitcelfer) (rotcev lanigiro) (rotcev noitcelfer) = rotcev detcelfer

,**r** rotcev lanigiro eht ,detcelfer eb lliw hcihw ,rotcev eht gnillac nehW
,**r**ref rotcev detcelfer eht noitcelfer eht retfa rotcev eht
,noitcelfer fo sixa eht fo noitcerid eht otni stniop hcihw ,rotcev tinu eht dna
tcudorp hciwdnas eht ,**n** rotcev noitcelfer

rref = **n r n**

.dnuof eb ot sah

elpmaxe na sA
,2 e_1 + e_3 = 7 e_1 + 5 e_2 + 6 e_3 = **r** rotcev eht tcelfer won lliw ew
,3 retpahc ni dessucsid neeb ydaerla sah hcihw

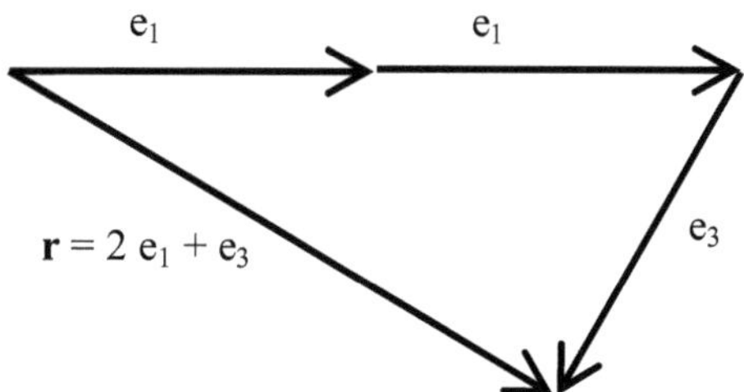

.3 e_1 + e_2 = **a** rotcev fo noitcerid eht otni stniop hcihw ,sixa na ta

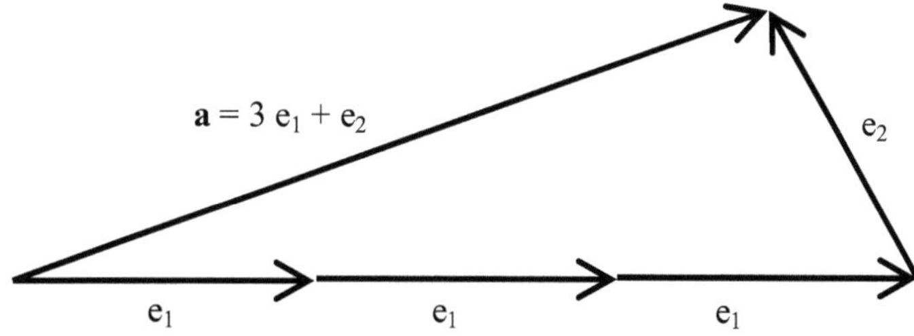

.noitcelfer fo sixa eht otni stniop hcihw ,rotcev tinu sa rotcev noitcelfer ehT

.tsrif dnuof eb ot sah

$$\mathbf{a} = 3\,e_1 + e_2 \quad\Rightarrow\quad \mathbf{a}^2 = (3\,e_1 + e_2)^2$$

$$= (3\,e_1 + e_2)\,(3\,e_1 + e_2)$$

$$= 9\,e_1{}^2 + 3\,e_1 e_2 + 3\,e_2 e_1 + e_2{}^2$$

$$= 10 + 3\,e_{12} + 3\,e_{21}$$

$$= 7 + \underbrace{3 + 3\,e_{12} + 3\,e_{21}}_{0}$$

$$= 7$$

$$2.5^2 + \sin^2 60° = 6.25 + 0.75 = 7 \qquad\qquad \text{:kcehc naerogahtyP}$$

,noitcelfer fo sixa eht fo noitcerid eht otni stniop hcihw ,**a** rotcev eht eroferehT

:thgnel gniwollof eht sah

$$a = \left|\mathbf{a}\right| = \left|3\,e_1 + e_2\right| = \sqrt{(3\,e_1 + e_2)^2} = \sqrt{7}$$

eb lliw sixa-**a** eht fo noitcerid eht otni rotcev tinu sa rotcev noitcelfer eht suhT

$$\mathbf{n} = \frac{\mathbf{a}}{\left|\mathbf{a}\right|} = \frac{1}{\sqrt{7}}\,(3\,e_1 + e_2)$$

:si neht tcudorp hciwdnas eht fo noitaluclac ehT

$$\mathbf{r}_{ref} = \mathbf{n}\,\mathbf{r}\,\mathbf{n}$$

$$= \frac{1}{\sqrt{7}}\,(3\,e_1 + e_2)\,(2\,e_1 + e_3)\,\frac{1}{\sqrt{7}}\,(3\,e_1 + e_2)$$

$$= \frac{1}{7}\,(6\,e_1{}^2 + 3\,e_1 e_3 + 2\,e_2 e_1 + e_2 e_3)\,(3\,e_1 + e_2)$$

$$= \frac{1}{7}\,(6 + 3\,e_{21} + 2\,e_{21} + e_{12})\,(3\,e_1 + e_2)$$

$$= \frac{1}{7}\,(5 + 4\,e_{21})\,(3\,e_1 + e_2)$$

$$= \frac{1}{7}\,(15\,e_1 + 5\,e_2 + 12\,e_{21}\,e_1 + 4\,e_{21}\,e_2)$$

teg ew , $e_3 = e_2\, e_1\, e_2 = e_{21}\, e_2$ noitcelfer eht dna $e_2 = e_2\, e_1\, e_1 = e_{21}\, e_1$ sA

$$\mathbf{r_{ref}} = \frac{1}{7}\,(15\, e_1 + 5\, e_2 + 12\, e_2 + 4\, e_3)$$

$$= \frac{1}{7}\,(15\, e_1 + 17\, e_2 + 4\, e_3)$$

$$= \frac{1}{7}\,(11\, e_1 + 13\, e_2) = \frac{11}{7}\, e_1 + \frac{13}{7}\, e_2 \approx 1.57\, e_1 + 1.86\, e_2$$

:noitcelfer siht swohs erugif gniwollof ehT

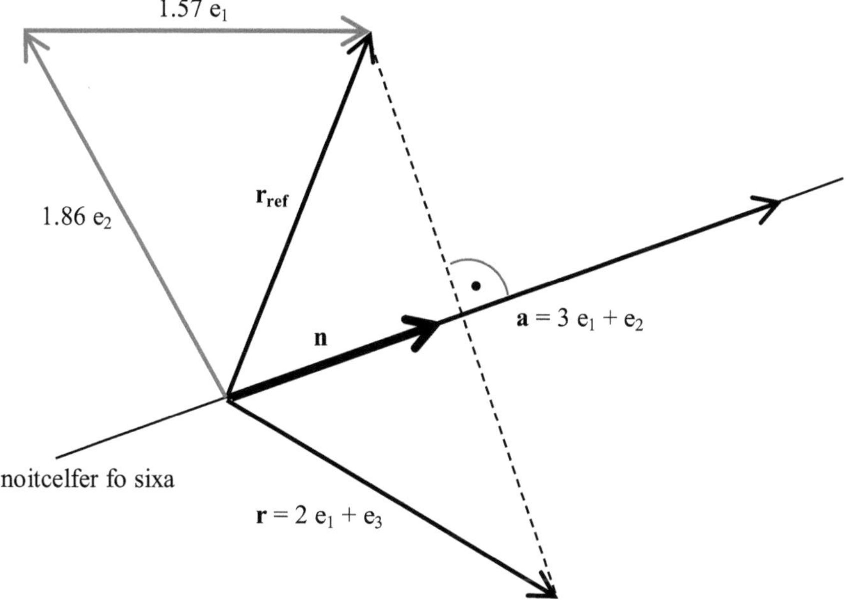

suoivbo si tI

noitcerid etisoppo ni $(e_{12} + e_{21})\,\mathbf{n} = \mathbf{u}$ rotcev noitcelfer a gnisu yb taht

$$\mathbf{u} = (e_{12} + e_{21})\,\mathbf{n} = (e_{12} + e_{21})\,\frac{1}{\sqrt{7}}\,(3\, e_1 + e_2) = \frac{1}{\sqrt{7}}\,(2\, e_2 + 3\, e_3) \qquad \text{with} \qquad \mathbf{u}^2 = 1$$

:tluser lacitnedi yletelpmoc a teg lliw ew

$$\mathbf{r_{ref}} = \mathbf{u}\,\mathbf{r}\,\mathbf{u}$$

$$= \frac{1}{\sqrt{7}}\,(2\,e_2 + 3\,e_3)\,(2\,e_1 + e_3)\,\frac{1}{\sqrt{7}}\,(2\,e_2 + 3\,e_3)$$

$$= \frac{1}{7}\,(4\,e_{21} + 2\,e_{12} + 6\,e_{12} + 3\,e_3{}^2)\,(2\,e_2 + 3\,e_3)$$

$$= \frac{1}{7}\,(3 + 8\,e_{12} + 4\,e_{21})\,(2\,e_2 + 3\,e_3)$$

$$= \frac{1}{7}\,(5\,e_{12} + e_{21})\,(2\,e_2 + 3\,e_3)$$

$$= \frac{1}{7}\,(10\,e_1 + 15\,e_1 e_2 e_3 + 2\,e_3 + 3\,e_2 e_1 e_3)$$

$$e_2 = e_3 e_1 e_3 = e_1 e_2 e_1\ e_1 e_3\ = e_1 e_2\ 1\ e_3 = e_1 e_2 e_3 \quad \text{sA}$$
$$\text{:teg ew ,}\ e_1 = e_3 e_2 e_3 = e_2 e_1 e_2\ e_2 e_3\ = e_2 e_1\ 1\ e_3 = e_2 e_1 e_3 \quad \text{dna}$$

$$\mathbf{r_{ref}} = \frac{1}{7}\,(10\,e_1 + 15\,e_2 + 2\,e_3 + 3\,e_1)$$

$$= \frac{1}{7}\,(13\,e_1 + 15\,e_2 + 2\,e_3)$$

$$= \frac{1}{7}\,(11\,e_1 + 13\,e_2) = \frac{11}{7}\,e_1 + \frac{13}{7}\,e_2 \approx 1{,}57\,e_1 + 1{,}86\,e_2$$

htgnel eht fo erauqs eht fo kcehc a noitidda nI

$$\mathbf{r}^2 = (2\,e_1 + e_3)^2 = 4 + 2\,e_{21} + 2\,e_{12} + 1 = 3$$

$$\mathbf{r_{ref}}^2 = \frac{1}{49}\,(11\,e_1 + 13\,e_2)^2 = \frac{1}{49}\,(121 + 143\,e_{12} + 143\,e_{21} + 169) = \frac{147}{49} = 3$$

$$\mathbf{r}^2 = \mathbf{r_{ref}}^2 \qquad \Leftarrow$$

.tcerroc si (ylekil tsom) tluser eht taht ,smrifnoc

si ti esruoc fo dnA

$$\mathbf{r} + \mathbf{r_{ref}} = 2\,e_1 + e_3 + \frac{11}{7}\,e_1 + \frac{13}{7}\,e_2 = \frac{25}{7}\,e_1 + \frac{13}{7}\,e_2 + e_3 = \frac{18}{7}\,e_1 + \frac{6}{7}\,e_2$$

$$= \frac{6}{7}\,(3\,e_1 + e_2) = \frac{6}{7}\,\mathbf{n}$$

$\mathbf{r_{ref}}$ rotcev detcelfer eht dna $\mathbf{r}$ rotcev lanigiro eht fo mus ehT
,rotcev noitcelfer eht fo elpitlum a eb tsum

because this vectorial sum must be parallel to the axis of reflection.

The parallel component $r_\parallel$ of both vectors $\mathbf{r}$ and $\mathbf{r_{ref}}$ therefore are:

$$\mathbf{r_\parallel} = \frac{1}{2}(\mathbf{r} + \mathbf{r_{ref}}) = \frac{9}{7}e_1 + \frac{3}{7}e_2 = \frac{3}{7}\mathbf{n}$$

And what about the difference of both vectors?

Negative numbers do not exist.
Minus signs do not exist.
$\mathbf{r} - \mathbf{r_{ref}}$ does not exist.

We will find this difference by calculating

$$\mathbf{r} \text{ minus } \mathbf{r_{ref}} = \mathbf{r} + (e_{12} + e_{21})\,\mathbf{r_{ref}}$$

instead.

Everything is positive! Always!
Minus signs are always, immediately and instantaneously replaced
by the positive sum of the two rhombuses $e_{21} + e_{12}$.

$$\mathbf{r} + (e_{12} + e_{21})\,\mathbf{r_{ref}} = 2\,e_1 + e_3 + \frac{1}{7}(e_{12} + e_{21})(11\,e_1 + 13\,e_2)$$

$$= 2\,e_1 + e_3 + \frac{11}{7}e_3 + \frac{13}{7}e_1 + \frac{11}{7}e_2 + \frac{13}{7}e_3$$

$$= \frac{27}{7}e_1 + \frac{11}{7}e_2 + \frac{31}{7}e_3$$

$$= \frac{16}{7}e_1 + \frac{20}{7}e_3$$

The orthogonal component $r_\perp$ of the vector $\mathbf{r}$ therefore is:

$$\mathbf{r_\perp} = \frac{1}{2}(\mathbf{r} + (e_{12} + e_{21})\,\mathbf{r_{ref}}) = \frac{8}{7}e_1 + \frac{10}{7}e_3$$

Adding both components $r_\parallel$ and $r_\perp$ will indeed result in the original vector $\mathbf{r}$:

$$\mathbf{r_\parallel} + \mathbf{r_\perp} = \frac{9}{7}e_1 + \frac{3}{7}e_2 + \frac{8}{7}e_1 + \frac{10}{7}e_3$$

$$= \frac{17}{7}e_1 + \frac{3}{7}e_2 + \frac{10}{7}e_3 = \frac{14}{7}e_1 + \frac{7}{7}e_3 = 2\,e_1 + e_3 = \mathbf{r}$$

$\mathbf{r}_\perp$ dna $\mathbf{r}_\|$ stnenopmoc htob fo mus-itna evitisop yllaer eht dnA
:$\mathbf{r}_{ref}$ rotcev detcelfer eht ni tluser deedni lliw

$$\mathbf{r}_\| + (e_{12} + e_{21})\, \mathbf{r}_\perp = \frac{9}{7} e_1 + \frac{3}{7} e_2 + \frac{8}{7} (e_{12} + e_{21})\, e_1 + \frac{10}{7} (e_{12} + e_{21})\, e_3$$

$$= \frac{9}{7} e_1 + \frac{3}{7} e_2 + \frac{8}{7} e_3 + \frac{8}{7} e_2 + \frac{10}{7} e_2 + \frac{10}{7} e_1$$

$$= \frac{19}{7} e_1 + \frac{21}{7} e_2 + \frac{8}{7} e_3$$

$$= \frac{11}{7} e_1 + \frac{13}{7} e_2 = \mathbf{r}_{ref}$$

egap tsal eht fo rotcev ecnereffid eht fo flah tcaxe eht si tuB
?$\mathbf{n}$ rotcev noitcelfer eht ot ralucidneprep yllaer
?tnenopmoc lanogohtro na yllaer $\mathbf{r}_\perp$ sI

.selgna tuoba erac ot emit tuoba si ti oS

selgnA   11

?n rotcev noitcelfer eht dna **r** rotcev lanigiro eht neewteb elgna eht si tahW
?**r**$_{\mathbf{ref}}$ rotcev detcelfer eht dna **n** rotcev noitcelfer eht neewteb elgna eht si tahw dnA

.edutingam emas eht evah dluohs selgna htob ,eruS
.tcudorp renni eht fo pleh eht htiw selgna htob fo edutingam eht dnif lliw ew dnA

:tod dlob ,taf ,kciht ,gib a htiw nettirw si **b** dna **a** srotcev owt fo tcudorp renni ehT

$$\mathbf{a} \bullet \mathbf{b} = \frac{1}{2}(\mathbf{a}\,\mathbf{b} + \mathbf{b}\,\mathbf{a})$$

.orez eb lliw tcudorp renni rieht ,rehto hcae ot ralucidneprep era srotcev owt fI

:elpmaxE
8/7 e$_1$ + 10/7 e$_3$ = $\mathbf{r}_\perp$ dna 9/7 e$_1$ + 3/7 e$_2$ = $\mathbf{r}_\|$ stnenopmoc owt ehT
tcudorp renni rieht exuaceb ,rehto hcae ot ralucidneprep era

$$\mathbf{r}_\| \bullet \mathbf{r}_\perp = \frac{1}{2}(\mathbf{r}_\|\,\mathbf{r}_\perp + \mathbf{r}_\perp\,\mathbf{r}_\|)$$

$$= \frac{1}{98}\big((9\,e_1 + 3\,e_2)(8\,e_1 + 10\,e_3) + (8\,e_1 + 10\,e_3)(9\,e_1 + 3\,e_2)\big)$$

$$= \frac{1}{98}(72 + 90\,e_{21} + 24\,e_{21} + 30\,e_{12} + 72 + 24\,e_{12} + 90\,e_{12} + 30\,e_{21})$$

$$= \frac{1}{98}(144 + 144\,e_{12} + 144\,e_{21})$$

$$= \frac{144}{98}(1 + e_{12} + e_{21})$$

$$= 0$$

.sraeppasid dna orez si

:dloh lliw noitaler gniwollof eht ,raeppasid ton seod tcudorp renni eht fI

srotcev tinu owt fo tcudorp renni ehT
.srotcev owt eseht neewteb elgna eht fo enisoc eht ot lacitnedi si

snoitseuq eht rewsna won lliw ew ,noitaler siht gnisu yB
.retpahc siht fo gninnigeb eht ta deksa α_2 dna α_1 noitcelfer fo selgna eht tuoba

36

,**srotcev tinu** era tfel eht no tah a htiw srotvev gniwollof ehT
.shtgnel rieht yb dedivid neeb dah yeht esuaceb

$$\hat{\mathbf{r}} = \frac{1}{\sqrt{3}}\,(2\,e_1 + e_3) \qquad \Leftarrow \qquad \mathbf{r} = 2\,e_1 + e_3$$

$$\hat{\mathbf{r}}_{\mathbf{ref}} = \frac{1}{7\sqrt{3}}\,(11\,e_1 + 13\,e_2) \qquad \Leftarrow \qquad \mathbf{r_{ref}} = \frac{1}{7}\,(11\,e_1 + 13\,e_2)$$

$$\mathbf{n} = \frac{1}{\sqrt{7}}\,(3\,e_1 + e_2) \qquad \Leftarrow \qquad \mathbf{a} = 3\,e_1 + e_2$$

:teg ew oS

$$\cos \alpha_1 = \hat{\mathbf{r}} \bullet \mathbf{n} = \frac{1}{2}\,(\hat{\mathbf{r}}\,\mathbf{n} + \mathbf{n}\,\hat{\mathbf{r}})$$

$$= \frac{1}{2\sqrt{21}}\,((2\,e_1 + e_3)\,(3\,e_1 + e_2) + (3\,e_1 + e_2)\,(2\,e_1 + e_3))$$

$$= \frac{1}{2\sqrt{21}}\,(6 + 2\,e_{12} + 3\,e_{12} + e_{21} + 6 + 3\,e_{21} + 2\,e_{21} + e_{12})$$

$$= \frac{1}{2\sqrt{21}}\,(12 + 6\,e_{12} + 6\,e_{21})$$

$$= \frac{6}{2\sqrt{21}} = \frac{3}{\sqrt{21}} = \sqrt{\frac{3}{7}} \approx 0.6547 \qquad \Rightarrow \qquad \alpha_1 = 49.11°$$

49.11° si **n** rotcev noitcelfer eht dna **r** rotcev lanigiro eht neewteb α_1 elgna ehT

:swohs noitaluclac ralimis a dnA

$$\cos \alpha_2 = \mathbf{n} \bullet \hat{\mathbf{r}}_{\mathbf{ref}} = \frac{1}{2}\,(\mathbf{n}\,\hat{\mathbf{r}}_{\mathbf{ref}} + \hat{\mathbf{r}}_{\mathbf{ref}}\,\mathbf{n})$$

$$= \frac{1}{14\sqrt{21}}\,((3\,e_1 + e_2)\,(11\,e_1 + 13\,e_2) + (11\,e_1 + 13\,e_2)\,(3\,e_1 + e_2))$$

$$= \frac{1}{14\sqrt{21}}\,(33 + 39\,e_{12} + 11\,e_{21} + 13 + 33 + 11\,e_{12} + 39\,e_{21} + 13)$$

$$= \frac{1}{14\sqrt{21}}\,(92 + 50\,e_{12} + 50\,e_{21})$$

$$= \frac{42}{14\sqrt{21}} = \frac{3}{\sqrt{21}} = \sqrt{\frac{3}{7}} \approx 0.6547 \qquad \Rightarrow \qquad \alpha_2 = 49.11°$$

49.11° si **r_ref** rotcev detcelfer eht dna **n** rotcev noitcelfer eht neewteb α_2 elgna ehT

⇒ Both angles have the same magnitude.

And we are now indeed able to calculate angles
(… as long as they are between 0° and 180° …)

And we can use this knowledge now to check the assumption of the introductory
chapter, that the angle between the tentacle arms of intelligent starfishes is 120°.

$$\cos \alpha_{\angle} = e_1 \bullet e_2 = e_2 \bullet e_3 = e_3 \bullet e_2 = \frac{1}{2}(e_1\,e_2 + e_2\,e_1) = \frac{1}{2}(e_{12} + e_{21})$$

The cosine of the tentacle arms can be compared
with the cosine of the interior angles α_Δ of an equilateral triangle.

$$\cos \alpha_\Delta = e_1 \bullet (e_1 + e_2) = e_2 \bullet (e_2 + e_3) = e_3 \bullet (e_3 + e_1)$$

$$= \frac{1}{2}(e_1(e_1 + e_2) + (e_1 + e_2)\,e_1) = \frac{1}{2}(e_1{}^2 + e_{12} + e_1{}^2 + e_{21})$$

$$= \frac{1}{2}(1 + e_{12} + e_{21} + 1) = \frac{1}{2}(0 + 1)$$

$$= \frac{1}{2} = 0.5 \qquad \Rightarrow \qquad \alpha_\Delta = 60°$$

Trigonometrically doubling this interior angle α_Δ of an equilateral triangle

First try: $\cos(2\,\alpha_\Delta) = 2\cos^2\alpha_\Delta - 1$ does not exist !!!

$$= 2\cos^2\alpha_\Delta + 0 - 1$$

$$= 2\cos^2\alpha_\Delta + 1 + e_{12} + e_{21} - 1$$

Second try: $\cos(2\,\alpha_\Delta) = 2\cos^2\alpha_\Delta + e_{12} + e_{21}$ does exist

and substituting $\cos\alpha_\Delta = 0.5$ results in:

$$\cos(2\,\alpha_\Delta) = 2 \cdot \frac{1}{2^2} + e_{12} + e_{21} = \frac{1}{2} + e_{12} + e_{21} = \frac{1}{2} + \underbrace{\frac{1}{2}e_{12} + \frac{1}{2}e_{21} + \frac{1}{2}e_{12} + \frac{1}{2}e_{21}}_{0}$$

$$= \frac{1}{2}(e_{12} + e_{21})$$

$$\Rightarrow \qquad \cos\alpha_{\angle} = \cos(2\,\alpha_\Delta) \qquad \Rightarrow \qquad \alpha_{\angle} = 2\,\alpha_\Delta = 2 \cdot 60° = 120°$$

⇒ As shown the angle between the tentacle arms of intelligent starfishes is 120°

.tcudorp renni eht fo pleh eht htiw dnuof eb nac selgnA
.deriuqer si tcudorp retuo eht saera dnif oT

:sa denifed si tcudorp retuo ehT

$$a \wedge b = \frac{1}{2}(a\,b + (e_{12} + e_{21})\,b\,a)$$

eht ot pu sdda tcudorp retuo eht tcudorp renni eht htiw rehtegoT
:arbeglA cirtemoeG fo tcudorP cirtemoeG lanoitnevnoc ,etelpmoc

$$a\,b = \frac{1}{2}(a\,b + b\,a) + \frac{1}{2}(a\,b + (e_{12} + e_{21})\,b\,a) = a \bullet b + a \wedge b$$

stsilaedi yranret ybmap-ybman dna stsilatnemadnuf yrtemmys ykcinif dna yssuF
stcudorp eerht gniwollof eht redisnoc thgim

$$a \blacklozenge b = \frac{1}{3}(a\,b + \quad b\,a)$$

$$a \blacktriangle b = \frac{1}{3}(a\,b + e_{12}\,b\,a)$$

$$a \blacktriangledown b = \frac{1}{3}(a\,b + e_{21}\,b\,a)$$

$$a\,b = a \blacklozenge b + a \blacktriangle b + a \blacktriangledown b \qquad\qquad \Leftarrow$$

.su rehtob ton dluohs msicitehtsea na hcus tub ,lufrtuaeb erom sa
?syad eseht skrauq ykriuq deen ew od lleh eht yhW

sah won $A = |A|$ aera eht fo edutingam hcihw dna A aera detneiro hcihW
?egap gniwollof eht no niaga nwohs si hcihw ,e_{12} submohr eht

(erugif eht fo trap thgir ees) tnemecalpsid lellarap yB
eb ot sah tsuj aera eht fo edutingam eht taht yletaidemmi nees eb nac ti

$$A = |e_1|\,\sin 60° = 1 \cdot \sin 60° = \sin 60° = \frac{1}{2}\sqrt{3} \approx 0.8660$$

.$\sin 60° = h$ si ylraelc submohr eht fo thgih eht esuaceb

.tcudorp retuo eht fo pleh eht htiw decudorper eb lliw tluser siht gniwollof eht nI

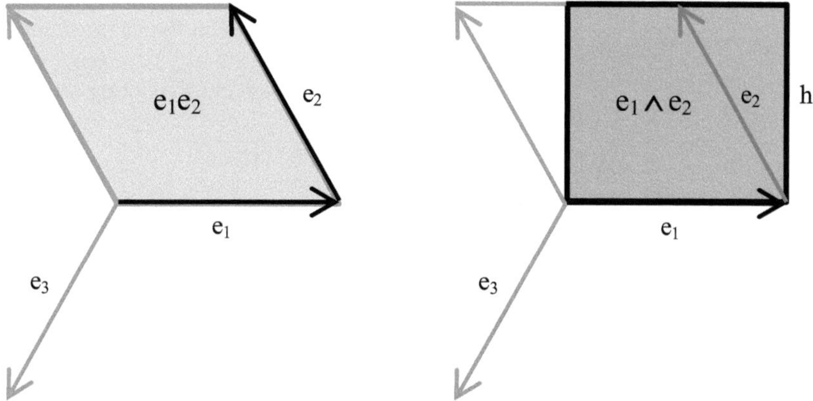

With $a = e_1$ and $b = e_2$ the outer product will be:

$$a \wedge b = \frac{1}{2}\left(a\,b + (e_{12} + e_{21})\,b\,a\right) = \frac{1}{2}\left(e_1\,e_2 + (e_{12} + e_{21})\,e_2\,e_1\right) = e_1 \wedge e_2$$

$$= \frac{1}{2}\left(e_{12} + 1 + e_{12}\right)$$

$$= \frac{1}{2} + e_{12}$$

This outer product describes the oriented area $A = e_1 \wedge e_2$ of the rhombus $e_1 e_2 = e_{12}$ shown above.

And please, mind the basis of Geometric Algebra: The outer product is an imaginary quantity!

The magnitude of this outer product can be found by a modified square with vectorial factors in reversed order.

Because of the distributing imaginarity it is not possible to get a square root of $(a \wedge b)^2$, but we have to calculate the square root of the product $(a \wedge b)(b \wedge a)$.

The second outer product thus has a reversed order:

$$b \wedge a = \frac{1}{2}\left(b\,a + (e_{12} + e_{21})\,a\,b\right) = \frac{1}{2}\left(e_2\,e_1 + (e_{12} + e_{21})\,e_1\,e_1\right) = e_2 \wedge e_1 = \frac{1}{2} + e_{21}$$

noitagujnoc xelpmoc a fo tros si erudecorp siht taht ezingocer thgim sdnim evitaerC
.sehsifrats tnegilletni fo csitamehtam eht ni

aera eht fo edutingam eht suhT

$$A = |A| = |a \wedge b| = \sqrt{(a \wedge b)(b \wedge a)}$$

eb lliw e₁₂ submohr eht fo

$$A = |\,e_1 \wedge e_2\,| = |\tfrac{1}{2} + e_{12}| = \sqrt{\left(\tfrac{1}{2} + e_{12}\right)\left(\tfrac{1}{2} + e_{21}\right)}$$

$$= \sqrt{\tfrac{1}{4} + \tfrac{1}{2}e_{21} + \tfrac{1}{2}e_{12} + 1} = \sqrt{\tfrac{5}{4} + \tfrac{1}{2}e_{21} + \tfrac{1}{2}e_{12}}$$

$$= \sqrt{\tfrac{3}{4}} = \tfrac{1}{2}\sqrt{3} \approx 0.8660$$

.tluser detcepxe eht deedni si sihT

stcudorp retuo dna renni fo scitamehtam ehT
.[6] ,[5] nnamssarG nnamreH yb depoleved neeb sah

pael mutnauq lacitamehtam citsatnaf dna tnecifingam ,dnarg sihT
.arbeglA cirtemoeG fo snoitadnuof lanoitpecnoc eht sebircsed

srotcev owt fo tcudorp retuo rehtona etaluclac lliw ew eroferehT
.eromyna srotcev tinu on era won hcihw

:melborp wen eht si ereH

margolellarap eht fo aera eht fo edutingam eht si tahW
? 3 e₁ + 2 e₂ = **b** dna 3 e₁ + e₂ = **a** srotcev edis owt eht htiw

(egap txen eht no nwohs) margolellarap sihT
fo esuaceb sa ,submohr a osla si

$$\mathbf{a}^2 = (3\,e_1 + e_2)^2 = 9 + 3\,e_{12} + 3\,e_{21} + 1 = 10 + 3\,e_{12} + 3\,e_{21} = 7$$

$$\mathbf{b}^2 = (3\,e_1 + 2\,e_2)^2 = 9 + 6\,e_{12} + 6\,e_{21} + 4 = 13 + 6\,e_{12} + 6\,e_{21} = 7$$

$$\sqrt{7} = |\,\mathbf{b}\,| = b = |\,\mathbf{a}\,| = a \quad \text{fo htgnel lacitnedi na evah srotcev edis htob}$$

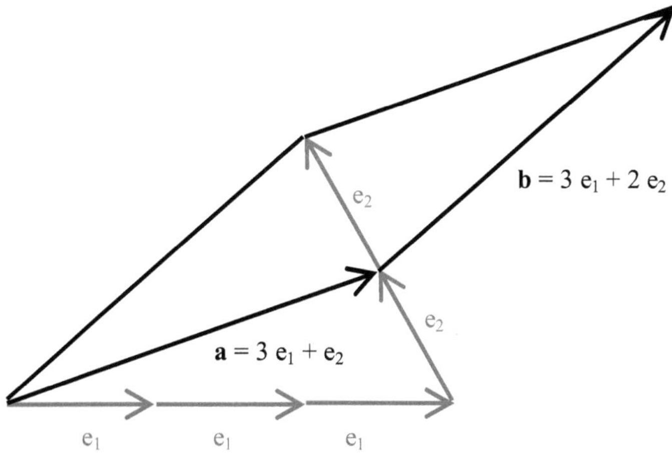

:era neht (submohr desrever osla dellac) submohr degnahc eht dna submohr ehT

$$\mathbf{b\,a} = (3\,e_1 + 2\,e_2)\,(3\,e_1 + e_2) \qquad\qquad \mathbf{a\,b} = (3\,e_1 + e_2)\,(3\,e_1 + 2\,e_2)$$

$$= 9 + 3\,e_{12} + 6\,e_{21} + 2 \qquad\qquad\qquad = 9 + 6\,e_{12} + 3\,e_{21} + 2$$

$$= 11 + 3\,e_{12} + 6\,e_{21} \qquad\qquad\qquad = 11 + 6\,e_{12} + 3\,e_{21}$$

$$= 8 + 3\,e_{21} \qquad\qquad\qquad\qquad = 8 + 3\,e_{12}$$

:eb neht lliw tcudorp retuo ehT

$$\mathbf{a} \wedge \mathbf{b} = \frac{1}{2}\,(\mathbf{a\,b} + (e_{12} + e_{21})\,\mathbf{b\,a}) = \frac{1}{2}\,(8 + 3\,e_{12} + (e_{12} + e_{21})\,(8 + 3\,e_{21}))$$

$$= \frac{1}{2}\,(8 + 3\,e_{12} + 8\,e_{12} + 3 + 8\,e_{21} + 3\,e_{12})$$

$$= \frac{1}{2}\,(11 + 14\,e_{12} + 8\,e_{21})$$

$$= \frac{1}{2}\,(3 + 6\,e_{12})$$

$$= 1.5 + 3\,e_{12}$$

:si neht submohr eht fo (srotcaf lairotcev eht fo redro eht gnignahc yb) noisrever ehT

$$\mathbf{b} \wedge \mathbf{a} = \frac{1}{2}\,(\mathbf{b\,a} + (e_{12} + e_{21})\,\mathbf{a\,b}) = 1.5 + 3\,e_{21}$$

Thus the following magnitude of the area of the rhombus can be found:

$$A = |\mathbf{A}| = |\mathbf{a} \wedge \mathbf{b}| \qquad = \sqrt{(\mathbf{a} \wedge \mathbf{b})(\mathbf{b} \wedge \mathbf{a})}$$

$$= |1.5 + 3\,e_{12}| = \sqrt{(1.5 + 3\,e_{12})(1.5 + 3\,e_{21})}$$

$$= \sqrt{2.25 + 4.5\,e_{21} + 4.5\,e_{12} + 9}$$

$$= \sqrt{11.25 + 4.5\,e_{21} + 4.5\,e_{12}}$$

$$= \sqrt{6.75} = 1.5\sqrt{3} \approx 2.5981$$

Hello mankind! Hello frantic lovers of negative signs!

The calculation just made is surely not more complicated than the usual calculation of standard Geometric Algebra with minus signs and negative values.

Here it is, the usual calculation, just to compare and to check the results. We recall standard Geometric Algebra [2], [3], [4], [8], [9], [10], [11], [12]:

The base vectors σ_x (unit vector to the right) and σ_y (upwards pointing unit vector) of standard Geometric Algebra will give the following representation of unit vectors of intelligent starfishes.

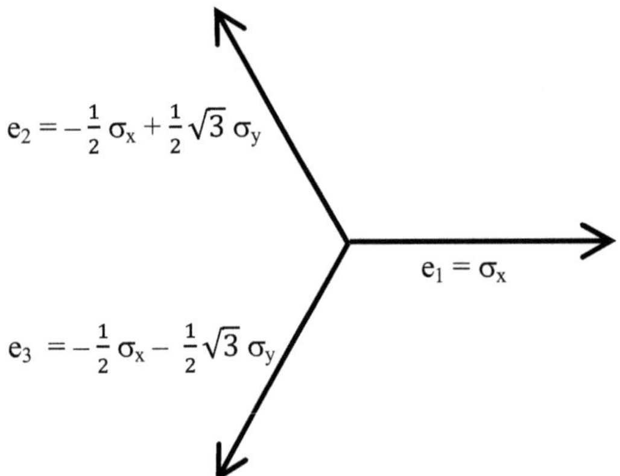

Both base vectors σ_x and σ_y then follow the basic equations of two-dimensional Pauli-Algebra:

$$\sigma_x \sigma_y = -\,\sigma_y \sigma_x \qquad \text{dna} \qquad \sigma_x{}^2 = \sigma_y{}^2 = 1$$

:era erofereht **b** dna **a** srotcev owt ehT

$$\mathbf{a} = 3\,e_1 + e_2 = 3\,\sigma_x - 0.5\,\sigma_x + 0.5\,\sqrt{3}\,\sigma_y = 2.5\,\sigma_x + 0.5\,\sqrt{3}\,\sigma_y$$

$$\mathbf{b} = 3\,e_1 + 2\,e_2 = 3\,\sigma_x + 2\,(-0.5\,\sigma_x + 0.5\,\sqrt{3}\,\sigma_y) = 2\,\sigma_x + \sqrt{3}\,\sigma_y$$

$$\mathbf{a}^2 = (2.5\,\sigma_x + 0.5\,\sqrt{3}\,\sigma_y)^2 = 6.25 + 0.75 = 7 \qquad \text{:kcehC}$$

$$\mathbf{b}^2 = (2\,\sigma_x + \sqrt{3}\,\sigma_y)^2 = 4 + 3 = 7$$

:submohr desrever dna submohR

$$\mathbf{a}\,\mathbf{b} = (2.5\,\sigma_x + 0.5\,\sqrt{3}\,\sigma_y)\,(2\,\sigma_x + \sqrt{3}\,\sigma_y) = 6.5 + 1.5\,\sqrt{3}\,\sigma_x \sigma_y$$

$$\mathbf{b}\,\mathbf{a} = (2\,\sigma_x + \sqrt{3}\,\sigma_y)\,(2.5\,\sigma_x + 0.5\,\sqrt{3}\,\sigma_y) = 6.5 - 1.5\,\sqrt{3}\,\sigma_x \sigma_y$$

:stcudorp retuO

$$\mathbf{a}\wedge\mathbf{b} = \frac{1}{2}\,(\mathbf{a}\,\mathbf{b} - \mathbf{b}\,\mathbf{a}) = \;\;1.5\,\sqrt{3}\,\sigma_x \sigma_y$$

$$\mathbf{b}\wedge\mathbf{a} = \frac{1}{2}\,(\mathbf{b}\,\mathbf{a} - \mathbf{a}\,\mathbf{b}) = -\,1.5\,\sqrt{3}\,\sigma_x \sigma_y$$

:submohr eht fo aera eht fo edutingaM

$$A = \left|\mathbf{A}\right| = \left|\mathbf{a}\wedge\mathbf{b}\right| = \sqrt{(\mathbf{a}\wedge\mathbf{b})\,(\mathbf{b}\wedge\mathbf{a})} = \left|\,1.5\,\sqrt{3}\,\sigma_x \sigma_y\right|$$

$$= 1.5\,\sqrt{3}$$

$$= \sqrt{6.75} \approx 2.5981 \qquad \Rightarrow \qquad \text{o.k.}$$

.lacitnedi era stluser htob esruoc fO :noisulcnoC

$\Leftarrow$ The parallelogram with the two side vectors $\mathbf{a} =$ 3 e₁ + e₂ dna $\mathbf{b} =$ 3 e₁ + 2 e₂ — as printed: 3 e_1 + 2 e_2 = **b** dna 3 e_1 + e_2 = **a** srotcev edis owt eht htiw margolellarap ehT $\Leftarrow$
.aera fo stinu 1.5 $\sqrt{3}$ fo aera na sah

.snoitator eurt rof emoc sah emit ,melborp siht devlos yletelpmoc won gnivah retfA
srotcev nwonk-llew ydaerla eht gnisuer yb delledom eb lliw snoitator eseht dnA
.retpahc gniwollof eht ni **b** dna **a**

13 Rotations

Rotations rotate. Reflections reflect.
Rotations and reflections are connected in a geometric fundamentally
and very very basic way:

Every rotation can be split into two reflections.
Mathematically rotations are composed of reflections.

Rotations are double reflections.
If we want to rotate an object, we only have to reflect it twice.

We will do this now with the vector $\mathbf{r} = 2\,e_1 + e_3$

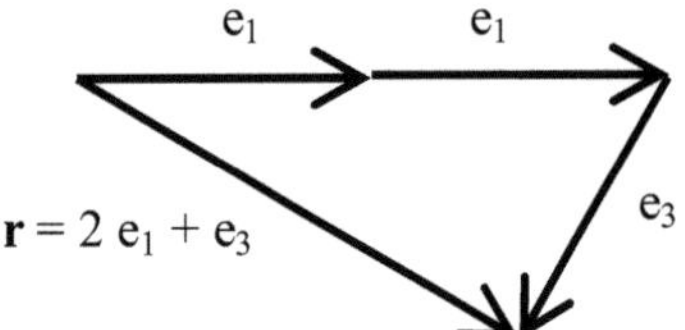

which we already know (and which is shown here with a smaller scale again).

Problem:

Which rotation do we get, if the vector $\mathbf{r} = 2\,e_1 + e_3$
first is reflected in an axis
which points into the direction of the vector $\mathbf{b} = 3\,e_1 + 2\,e_2$
and after that the reflected vector is reflected in a second axis
which points into the direction of the vector $\mathbf{a} = 3\,e_1 + e_2$?

The geometric situation at the beginning
is shown in the figure on the following page.

The reflection vector of the first reflection is the unit vector

$$\mathbf{m} = \frac{\mathbf{b}}{|\mathbf{b}|} = \frac{1}{\sqrt{7}}\,(3\,e_1 + 2\,e_2)$$

Thus the following first sandwich product has to be found:

$$\mathbf{r}_{ref} = \mathbf{m}\,\mathbf{r}\,\mathbf{m}$$

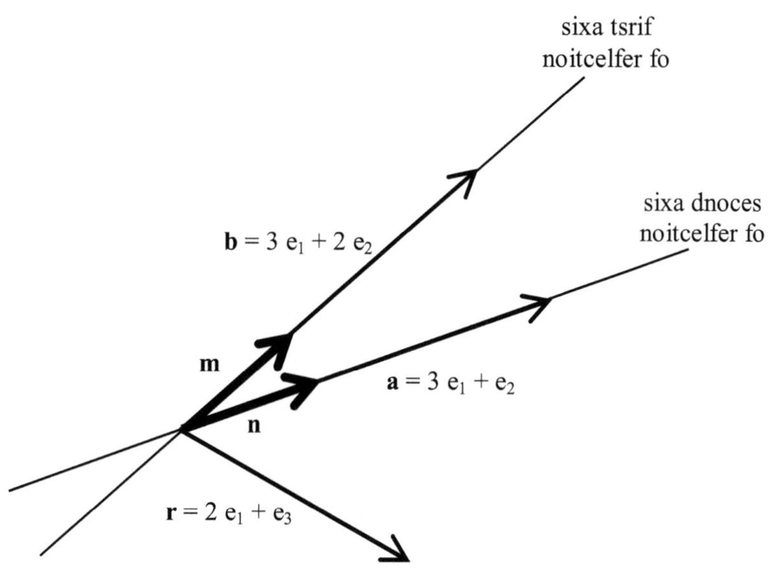

$$\Rightarrow \qquad \mathbf{r_{ref}} = \frac{1}{\sqrt{7}}\,(3\ e_1 + 2\ e_2)\,(2\ e_1 + e_3)\,\frac{1}{\sqrt{7}}\,(3\ e_1 + 2\ e_2)$$

$$= \frac{1}{7}\,(6 + 3\ e_{21} + 4\ e_{21} + 2\ e_{12})\,(3\ e_1 + 2\ e_2)$$

$$= \frac{1}{7}\,(4 + 5\ e_{21})\,(3\ e_1 + 2\ e_2)$$

$$= \frac{1}{7}\,(12\ e_1 + 8\ e_2 + 15\ e_2 + 10\ e_3)$$

$$= \frac{1}{7}\,(2\ e_1 + 13\ e_2) = \frac{2}{7}\ e_1 + \frac{13}{7}\ e_2 \approx 0.29\ e_1 + 1.86\ e_2$$

.egap gniwollof eht fo erugif eht ni nwohs si tluser etaidemretni sihT

:thtgnel eht fo kcehc a tsrif tuB

$$\mathbf{r_{ref}}^2 = \frac{1}{49}\,(2\ e_1 + 13\ e_2)^2 = \frac{1}{49}\,(4 + 26\ e_{12} + 26\ e_{21} + 169) = \frac{147}{49} = 3$$

:srotcev htob fo mus eht gnikcehc dnA

$$r + r_{ref} = 2\,e_1 + e_3 + \frac{2}{7}\,e_1 + \frac{13}{7}\,e_2 = \frac{16}{7}\,e_1 + \frac{13}{7}\,e_2 + \frac{7}{7}\,e_3 = \frac{9}{7}\,e_1 + \frac{6}{7}\,e_2 = \frac{3}{\sqrt{7}}\,m$$

.m fo elpitlum a si $r + r_{ref}$ $\Leftarrow$

rotcev noitcelfer dnoces ehT

$$n = \frac{a}{|a|} = \frac{1}{\sqrt{7}}\,(3\,e_1 + e_2)$$

,10 retpahc ni dnuof neeb ydaerla sah
tcudorp hciwdnas dnoces gniwollof eht gnivig suht

$$r_{rot} = n\,r_{ref}\,n$$

$$= \frac{1}{\sqrt{7}}\,(3\,e_1 + e_2)\,\frac{1}{7}\,(2\,e_1 + 13\,e_2)\,\frac{1}{\sqrt{7}}\,(3\,e_1 + e_2)$$

$$= \frac{1}{49}\,(6 + 39\,e_{12} + 2\,e_{21} + 13)\,(3\,e_1 + e_2)$$

$$= \frac{1}{49}\,(17 + 37\,e_{12})\,(3\,e_1 + e_2)$$

$$\mathbf{r_{rot}} = \frac{1}{49}\,(51\ e_1 + 17\ e_2 + 111\ e_3 + 37\ e_1)$$

$$= \frac{1}{49}\,(71\ e_1 + 94\ e_3) = \frac{71}{49}\ e_1 + \frac{94}{49}\ e_3 \approx 1.45\ e_1 + 1.92\ e_3$$

:erugif (lanoisivorp tub) laniF

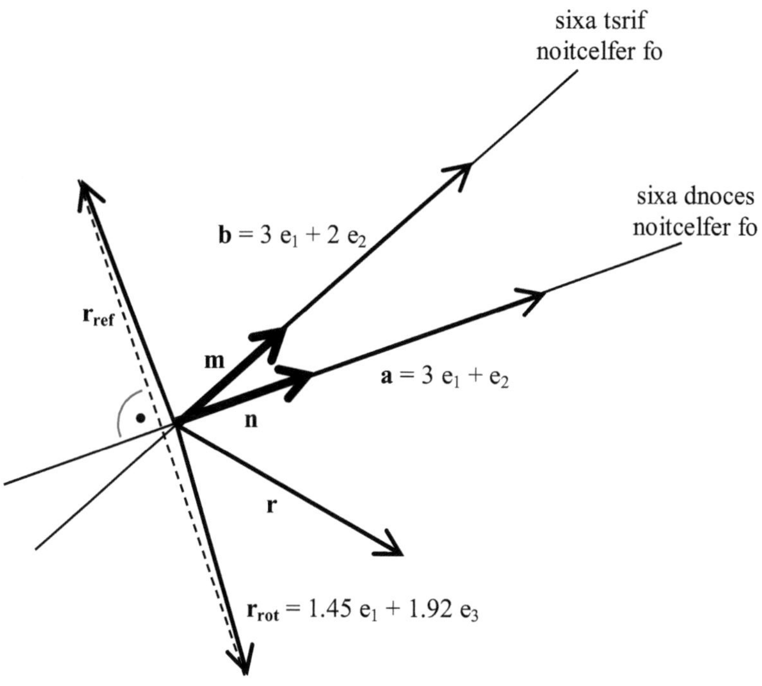

:htgnel eht gnikcehC

$$\mathbf{r_{rot}}^2 = \frac{1}{2401}\,(71\ e_1 + 94\ e_3)^2 = \frac{1}{2401}\,(5041 + 6674\ e_{21} + 6674\ e_{12} + 8836) = \frac{7203}{2401} = 3$$

:mus eht gnikcehC

$$\mathbf{r_{ref}} + \mathbf{r_{rot}} = \frac{2}{7}\ e_1 + \frac{13}{7}\ e_2 + \frac{71}{49}\ e_1 + \frac{94}{49}\ e_3 = \frac{85}{49}\ e_1 + \frac{91}{49}\ e_2 + \frac{94}{49}\ e_3 = \frac{6}{49}\ e_2 + \frac{9}{49}\ e_3$$

$$= (e_{12} + e_{21})^2 \left(\frac{6}{49}\ e_2 + \frac{9}{49}\ e_3\right) = (e_{12} + e_{21})\left(\frac{6}{49}\ e_1 + \frac{9}{49}\ e_2 + \frac{6}{49}\ e_3 + \frac{9}{49}\ e_1\right)$$

$$\mathbf{r}_{ref} + \mathbf{r}_{rot} = (e_{12} + e_{21})\left(\frac{9}{49}\,e_1 + \frac{3}{49}\,e_2\right) = \frac{3}{7\sqrt{7}}\,(e_{12} + e_{21})\,\mathbf{n}$$

.**n** fo elpitlum a si $\mathbf{r}_{ref} + \mathbf{r}_{rot}$ ⇐

:noisulcnoc elbanoitseuQ

71/49 e_1 + 94/49 e_3 = $\mathbf{r}_{rot}$ rotcev eht otni detator si 2 e_1 + e_3 = **r** rotcev ehT
.snoitcelfer owt eht yb

?noitator a yllautca siht si tuB
.dlot era ew tahw ylevïan eveileb reven dluohs eW
!snoitressa suoibud kcehc ot esnes sekam syawla tI

noitcelfer a wohs ylbissop osla dluoc egap suoiverp eht fo erugif lanif ehT
:$\mathbf{r}_{rot}$ dna **r** neewteb elgna eht stcesib hcihw sixa na ta **r** rotcev lanigiro eht fo

$$\mathbf{c} = \frac{1}{2}\,(\mathbf{r} + \mathbf{r}_{rot}) = 2\,e_1 + e_3 + \frac{71}{49}\,e_1 + \frac{94}{49}\,e_3 = \frac{169}{49}\,e_1 + \frac{143}{49}\,e_3$$

:noitcelfer fo rotcev tinU

$$\text{ɯ} = \frac{\mathbf{c}}{|\mathbf{c}|} = \frac{1}{7\sqrt{3}}\,(13\,e_1 + 11\,e_3) \qquad \Rightarrow \qquad \text{ɯ}^2 = 1$$

:tcudorp hciwdnaS

$$\text{ɯ}\,\mathbf{r}\,\text{ɯ} = \frac{1}{7\sqrt{3}}\,(13\,e_1 + 11\,e_3)\,(2\,e_1 + e_3)\,\frac{1}{7\sqrt{3}}\,(13\,e_1 + 11\,e_3)$$

$$= \frac{1}{147}\,(24 + 9\,e_{12})\,(13\,e_1 + 11\,e_3) = \frac{1}{49}\,(8 + 3\,e_{12})\,(13\,e_1 + 11\,e_3)$$

$$= \frac{1}{49}\,(104\,e_1 + 33\,e_2 + 127\,e_3) = \frac{71}{49}\,e_1 + \frac{94}{49}\,e_3 = \mathbf{r}_{rot}$$

,noitator a si yletinifed siht taht ees ot elba eb oT
gel a sa 2 e_1 + e_3 = **r** rotcev .g.e ,deriuqer si tcejbo lacirtemmysa na
0.5 e_3 = **s** rotcev dnoces a fo noitcerid eht otni toof a htiw

:melborp deliated erom ,weN

2 e_1 + e_3 = **r** rotcev eht fi ,teg ew od noitator hcihW
0.5 e_3 = **s** rotcev toof lanoitidda eht htiw
sixa na ni detcelfer si tsrif
3 e_1 + 2 e_2 = **b** rotcev eht fo noitcerid eht otni stniop hcihw

sixa dnoces a ni dercelfer era srotcev detcelfer eht taht retfa dna
? $3\ e_1 + e_2 =$ **a** rotcev eht fo noitcerid eht otni stniop hcihw

noitseuq siht rewsna oT
erugif ees) seot eht ot stniop hcihw , $2\ e_1 + 1.5\ e_3 =$ **s** + **r** mus rotcev dettod eht
.**s** dna **r** srotcev eht fo snoitcelfer etarepes yb sexa owt eht ta detcelfer si ,(woleb

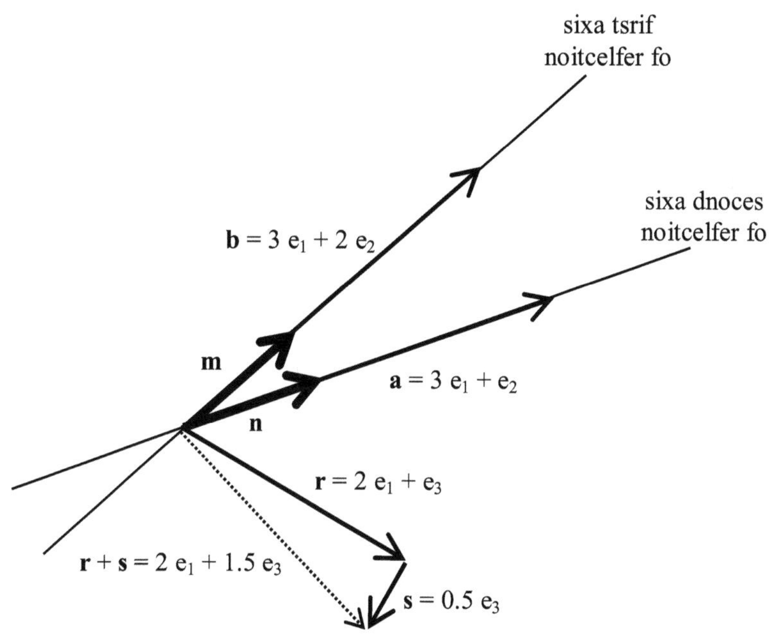

:**b** rotcev fo noitcerid eht otni stniop hcihw sixa eht ta noitcelfer tsriF

$$s_{ref} = \mathbf{m}\,\mathbf{s}\,\mathbf{m}$$

$$= \frac{1}{\sqrt{7}}\,(3\,e_1 + 2\,e_2)\,\frac{1}{2}\,e_3\,\frac{1}{\sqrt{7}}\,(3\,e_1 + 2\,e_2)$$

$$= \frac{1}{14}\,(2\,e_{12} + 3\,e_{21})\,(3\,e_1 + 2\,e_2)$$

$$= \frac{1}{14}\,(6\,e_3 + 4\,e_1 + 9\,e_2 + 6\,e_3)$$

$$s_{ref} = \frac{1}{14}\,(5\,e_2 + 8\,e_3) = \frac{5}{14}\,e_2 + \frac{8}{14}\,e_3 \approx 0.36\,e_2 + 0.57\,e_3$$

:htgnel eht gnikcehC

$$s_{ref}^2 = \frac{1}{196}\,(5\,e_2 + 8\,e_3)^2 = \frac{1}{196}\,(25 + 40\,e_{12} + 40\,e_{21} + 64) = \frac{49}{196} = \frac{1}{4} = s^2$$

:mus eht gnikcehC

$$s + s_{ref} = \frac{1}{2}\,e_3 + \frac{5}{14}\,e_2 + \frac{8}{14}\,e_3 = \frac{5}{14}\,e_2 + \frac{15}{14}\,e_3 = (e_{12} + e_{21})^2 \left(\frac{5}{14}\,e_2 + \frac{15}{14}\,e_3\right)$$

$$= (e_{12} + e_{21}) \left(\frac{15}{14}\,e_1 + \frac{10}{14}\,e_2\right) = (e_{12} + e_{21})\,\frac{5}{2\sqrt{7}}\,\mathbf{m}$$

$\Leftarrow$ .m fo elpitlum a si $s + s_{ref}$

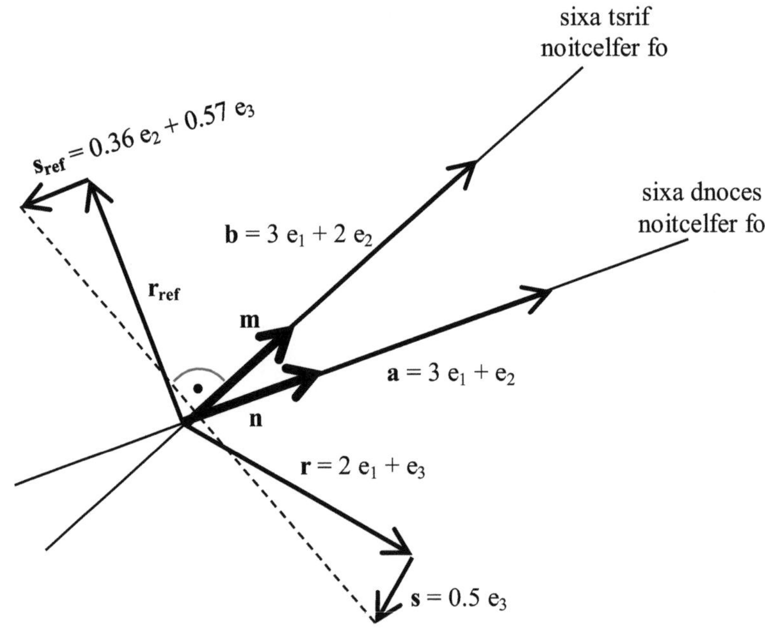

noitisop eht ta dnuof eb erofereht nac seot detcelfer ehT

$$r_{ref} + s_{ref} = \frac{2}{7}\,e_1 + \frac{13}{7}\,e_2 + \frac{5}{14}\,e_2 + \frac{8}{14}\,e_3 = \frac{27}{14}\,e_2 + \frac{2}{7}\,e_3 \approx 1.93\,e_2 + 0.29\,e_3$$

sixa eht ta noitcelfer dnoces eht htiw seunitnoc siht llA
:**a** rotcev fo noitcerid eht otni stniop hcihw

$$s_{rot} = \mathbf{n}\; s_{ref}\; \mathbf{n}$$

$$= \frac{1}{\sqrt{7}}\,(3\,e_1 + e_2)\,\frac{1}{14}\,(5\,e_2 + 8\,e_3)\,\frac{1}{\sqrt{7}}\,(3\,e_1 + e_2)$$

$$= \frac{1}{98}\,(18\,e_{12} + 19\,e_{21})\,(3\,e_1 + e_2)$$

$$= \frac{1}{98}\,(39\,e_2 + 55\,e_3) = \frac{39}{98}\,e_2 + \frac{55}{98}\,e_3 \approx 0.40\,e_2 + 0.56\,e_3$$

:htgnel eht gnikcehC

$$s_{rot}{}^2 = \frac{1}{9604}\,(39\,e_2 + 55\,e_3)^2 = \frac{1}{9604}\,(4546 + 2145\,e_{12} + 2145\,e_{21}) = \frac{2401}{9604} = \frac{1}{4} = s^2$$

:mus eht gnikcehC

$$s_{ref} + s_{rot} = \frac{5}{14}\,e_2 + \frac{8}{14}\,e_3 + \frac{39}{98}\,e_2 + \frac{55}{98}\,e_3 = \frac{74}{98}\,e_2 + \frac{111}{98}\,e_3$$

$$= (e_{12} + e_{21})^2 \left(\frac{74}{98}\,e_2 + \frac{111}{98}\,e_3\right) = (e_{12} + e_{21})\left(\frac{111}{98}\,e_1 + \frac{37}{98}\,e_2\right) = (e_{12} + e_{21})\,\frac{37}{14\sqrt{7}}\,\mathbf{n}$$

$\Leftarrow$ **.n** fo elpitlum a si $s_{ref} + s_{rot}$

noitisop eht ta dnuof eb eroferet nac seot detator ehT

$$r_{rot} + s_{roz} = \frac{71}{49}\,e_1 + \frac{94}{49}\,e_3 + \frac{39}{98}\,e_2 + \frac{55}{98}\,e_3 = \frac{103}{98}\,e_1 + \frac{102}{49}\,e_3 \approx 1.05\,e_1 + 2.08\,e_3$$

> **!noitator a si ylraelc siht dnA**

swohs ylsuougibmanu egap gniwollof eht no erugif ehT
ш rotcev gnitcesib eht ta noitcelfer a ton si ti taht
.toof eht fo edis lanigiro ,emas eht ot detneiro si gel eht esuaceb

.detator neeb evah ylraelc eroferet **s** rotcev htiw rehtegot **r** rotcev ehT

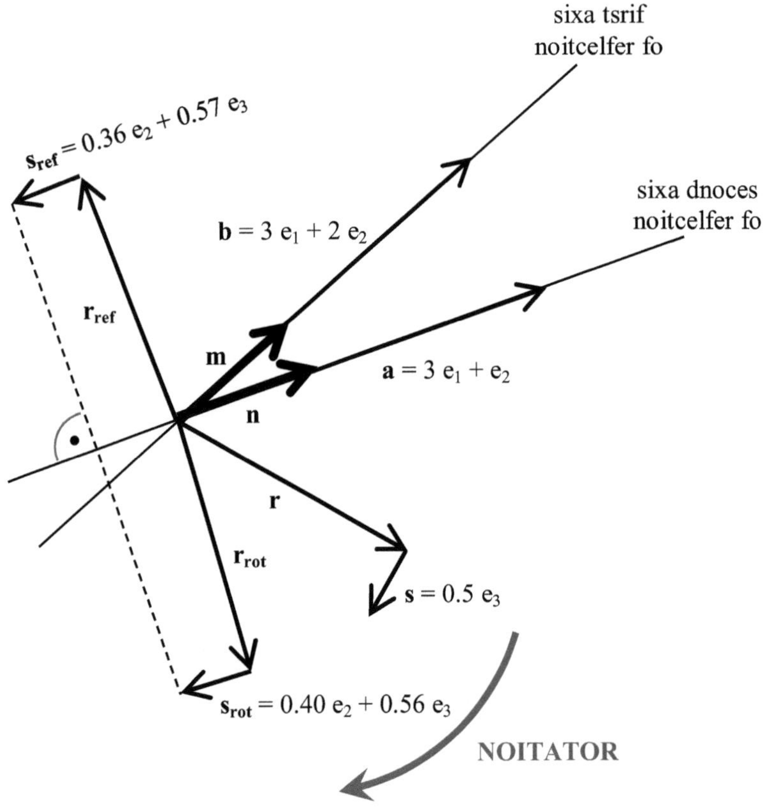

!sngis evitagen fo srevol citnarf olleH !dniknam olleH

.erehwyna raeppa sngis evitagen on snoitaluclac eseht lla nI

.hguorht dna hguorht evitisop era snoitarepo lla dna seititnauq llA

sngis evitagen htiw snoitaluclac lanoitnevnoc esruoc fo dnA
.stluser lcaitnedi yllatot ni tluser arbeglA cirtemoeG dradnats gnisu yb

.stluser ruo kcehc ot dna erapmoc ot tsuj ,noitaluclac lausu eht ,niaga si ti ereH

Short repetition:

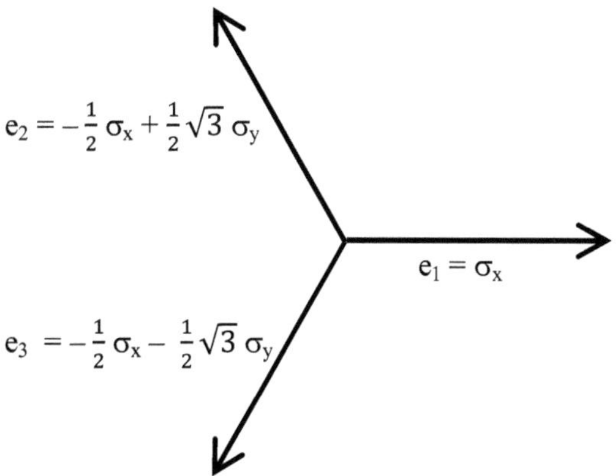

The base vectors σ_x (unit vector to the right)
and σ_y (unit vector pointing upwards)
follow the basic rules of two-dimensional Pauli-Algebra:

$$\sigma_x\sigma_y = -\,\sigma_y\sigma_x \qquad \text{and} \qquad \sigma_x^{\,2} = \sigma_y^{\,2} = 1$$

The two axes of reflection are pointing into the direction of the following vectors:

$$\mathbf{a} = 3\,e_1 + e_2 = 3\,\sigma_x - 0.5\,\sigma_x + 0.5\sqrt{3}\,\sigma_y = 2.5\,\sigma_x + 0.5\sqrt{3}\,\sigma_y$$

$$\mathbf{b} = 3\,e_1 + 2\,e_2 = 3\,\sigma_x + 2\,(-\,0.5\,\sigma_x + 0.5\sqrt{3}\,\sigma_y) = 2\,\sigma_x + \sqrt{3}\,\sigma_y$$

Reflection vectors:

$$\mathbf{n} = \frac{\mathbf{a}}{|\mathbf{a}|} = \frac{5}{2\sqrt{7}}\,\sigma_x + \frac{\sqrt{3}}{2\sqrt{7}}\,\sigma_y \qquad \Leftarrow \qquad \mathbf{a}^2 = 6.25 + 0.75 = 7$$

$$\mathbf{m} = \frac{\mathbf{b}}{|\mathbf{b}|} = \frac{2}{\sqrt{7}}\,\sigma_x + \frac{\sqrt{3}}{\sqrt{7}}\,\sigma_y \qquad \Leftarrow \qquad \mathbf{b}^2 = 4 + 3 = 7$$

Original vectors:

$$\mathbf{r} = 2\,e_1 + e_3 = 2\,\sigma_x - \frac{1}{2}\,\sigma_x - \frac{1}{2}\sqrt{3}\,\sigma_y = \frac{3}{2}\,\sigma_x - \frac{1}{2}\sqrt{3}\,\sigma_y$$

$$\mathbf{s} = \frac{1}{2}\,e_3 = -\frac{1}{4}\,\sigma_x - \frac{1}{4}\sqrt{3}\,\sigma_y$$

54

:stcudorp hciwdnaS

$$\mathbf{r_{ref}} = \mathbf{m\ r\ m} = \frac{1}{\sqrt{7}}\,(2\,\sigma_x + \sqrt{3}\,\sigma_y)\,\frac{1}{2}\,(3\,\sigma_x - \sqrt{3}\,\sigma_y)\,\frac{1}{\sqrt{7}}\,(2\,\sigma_x + \sqrt{3}\,\sigma_y)$$

$$= \frac{1}{14}\,(3 - 5\,\sqrt{3}\,\sigma_x\sigma_y)\,(2\,\sigma_x + \sqrt{3}\,\sigma_y)$$

$$= \frac{1}{14}\,(-9\,\sigma_x + 13\,\sqrt{3}\,\sigma_y) = \frac{1}{7}\,(2\,e_1 + 13\,e_2) \qquad \Rightarrow \qquad \text{o.k.}$$

$$\mathbf{r_{rot}} = \mathbf{n\ r_{ref}\ n} = \frac{1}{2\sqrt{7}}\,(5\,\sigma_x + \sqrt{3}\,\sigma_y)\,\frac{1}{14}\,(-9\,\sigma_x + 13\,\sqrt{3}\,\sigma_y)\,\frac{1}{2\sqrt{7}}\,(5\,\sigma_x + \sqrt{3}\,\sigma_y)$$

$$= \frac{1}{392}\,(-6 + 74\,\sqrt{3}\,\sigma_x\sigma_y)\,(5\,\sigma_x + \sqrt{3}\,\sigma_y)$$

$$= \frac{1}{49}\,(24\,\sigma_x - 47\,\sqrt{3}\,\sigma_y) = \frac{1}{49}\,(71\,e_1 + 94\,e_3) \qquad \Rightarrow \qquad \text{o.k.}$$

$$\mathbf{s_{ref}} = \mathbf{m\ s\ m} = \frac{1}{\sqrt{7}}\,(2\,\sigma_x + \sqrt{3}\,\sigma_y)\,\frac{-1}{4}\,(\sigma_x + \sqrt{3}\,\sigma_y)\,\frac{1}{\sqrt{7}}\,(2\,\sigma_x + \sqrt{3}\,\sigma_y)$$

$$= -\frac{1}{28}\,(5 + \sqrt{3}\,\sigma_x\sigma_y)\,(2\,\sigma_x + \sqrt{3}\,\sigma_y)$$

$$= -\frac{1}{28}\,(13\,\sigma_x + 3\,\sqrt{3}\,\sigma_y) = \frac{1}{14}\,(5\,e_2 + 8\,e_3) \qquad \Rightarrow \qquad \text{o.k.}$$

$$\mathbf{s_{rot}} = \mathbf{n\ s_{ref}\ n} = \frac{1}{2\sqrt{7}}\,(5\,\sigma_x + \sqrt{3}\,\sigma_y)\,\frac{-1}{28}\,(13\,\sigma_x + 3\,\sqrt{3}\,\sigma_y)\,\frac{1}{2\sqrt{7}}\,(5\,\sigma_x + \sqrt{3}\,\sigma_y)$$

$$= -\frac{1}{784}\,(74 + 2\,\sqrt{3}\,\sigma_x\sigma_y)\,(5\,\sigma_x + \sqrt{3}\,\sigma_y)$$

$$= -\frac{1}{98}\,(47\,\sigma_x + 8\,\sqrt{3}\,\sigma_y) = \frac{1}{98}\,(39\,e_2 + 55\,e_3) \qquad \Rightarrow \qquad \text{o.k.}$$

:seot detatoR

$$\mathbf{r_{rot}} + \mathbf{s_{roz}} = \frac{1}{98}\,(\sigma_x - 102\,\sqrt{3}\,\sigma_y) = \frac{1}{98}\,(103\,e_1 + 204\,e_3) \qquad \Rightarrow \qquad \text{o.k}$$

.stluser arbegla hsifrats eht htiw lacitnedi era stluser llA   ⇐
... noitator fo elgna eht fo ezis eht tuoba noitseuq eht si ereht llits tuB

srotoR dna noitatoR fo selgnA   14

,noitator fo elgna eht dnif oT
dessucsid neeb dah hcihw tcudorp renni eht fo seitreporp eht llacer ew
.11 retpahc ni

:α elgna gniwollof eht tuoba detator neeb sah **r** rotcev gel ehT

$$\cos \alpha = \hat{\mathbf{r}} \bullet \hat{\mathbf{r}}_{\mathbf{rot}} = \frac{1}{2}\,(\hat{\mathbf{r}}\,\hat{\mathbf{r}}_{\mathbf{rot}} + \hat{\mathbf{r}}_{\mathbf{rot}}\,\hat{\mathbf{r}})$$

$$= \frac{1}{2\cdot 3}\left((2\,e_1 + e_3)\,\frac{1}{49}\,(71\,e_1 + 94\,e_3) + \frac{1}{49}\,(71\,e_1 + 94\,e_3)\,(2\,e_1 + e_3)\right)$$

$$= \frac{1}{294}\,(142 + 188\,e_{21} + 71\,e_{12} + 94 + 142 + 71\,e_{21} + 188\,e_{12} + 94)$$

$$= \frac{1}{294}\,(472 + 259\,e_{12} + 259\,e_{21})$$

$$= \frac{213}{294} = \frac{71}{98} \approx 0.7245 \qquad \Rightarrow \qquad \alpha = 43.57°$$

tcudorp renni eht gnikat yb dnuof eb osla nac noitator fo elgna siht esruoc fO
… toof eht fo $\hat{\mathbf{s}}_{\mathbf{rot}}$ dna $\hat{\mathbf{s}}$ srotcev tinu detator dna lanigiro eht fo

$$\cos \alpha = \hat{\mathbf{s}} \bullet \hat{\mathbf{s}}_{\mathbf{rot}} = \frac{1}{2}\,(\hat{\mathbf{s}}\,\hat{\mathbf{s}}_{\mathbf{rot}} + \hat{\mathbf{s}}_{\mathbf{rot}}\,\hat{\mathbf{s}})$$

$$= \frac{1}{2}\left(e_3\,\frac{1}{49}\,(39\,e_2 + 55\,e_3) + \frac{1}{49}\,(39\,e_2 + 55\,e_3)\,e_3\right)$$

$$= \frac{1}{98}\,(55 + 39\,e_{12} + 39\,e_{21} + 55)$$

$$= \frac{71}{98} \approx 0.7245 \qquad \Rightarrow \qquad \alpha = 43.57°$$

snoitisop eht fo $\hat{\mathbf{t}}_{\mathbf{rot}}$ dna $\hat{\mathbf{t}}$ srotcev tinu detator dna lanigiro eht fo ro …
.seot eht fo

,thgnel rieht yb dedivid eb dluohs **r** + **s** = **t** srotcev noitisop eseht suhT

$$\frac{13}{4} = \left(2\,e_1 + \frac{3}{2}\,e_3\right)^2 = \mathbf{t}_{\mathbf{rot}}{}^2 = \mathbf{t}^2 \qquad \text{fo erauqs a sah hcihw}$$

:si neht α noitator fo elgna eht fo noitatupmoc evitanretla driht ehT

$$\cos\alpha = \hat{\mathbf{t}} \bullet \hat{\mathbf{t}}_{\mathbf{rot}} = \frac{1}{2}\,(\hat{\mathbf{t}}\,\hat{\mathbf{t}}_{\mathbf{rot}} + \hat{\mathbf{t}}_{\mathbf{rot}}\,\hat{\mathbf{t}})$$

$$= \frac{4}{2\cdot 13}\left(\left(2\,e_1 + \frac{3}{2}\,e_3\right)\frac{1}{98}\,(103\,e_1 + 204\,e_3) + \frac{1}{98}\,(103\,e_1 + 204\,e_3)\left(2\,e_1 + \frac{3}{2}\,e_3\right)\right)$$

$$= \frac{1}{1274}\,(412 + 816\,e_{21} + 309\,e_{12} + 612 + 412 + 309\,e_{21} + 816\,e_{12} + 612)$$

$$= \frac{1}{1274}\,(2048 + 1125\,e_{12} + 1125\,e_{21})$$

$$= \frac{923}{1274} = \frac{71}{98} \approx 0.7245 \qquad \Rightarrow \qquad \alpha = 43.57°$$

43.57° = α si noitator fo elgna eht fo edutingam ehT   ⇐

neewteb β elgna roiretni eht no sdneped α noitator fo elgna ehT
.(**m** dna **n** srotcev eht fo noitcerid eht otni tniop hcihw) noitcelfer fo sexa owt eht

:noitcelfer fo sexa eht neewteb β elgna roiretni siht etupmoc won lliw ew eroferehT

$$\cos\beta = \mathbf{n} \bullet \mathbf{m} = \frac{1}{2}\,(\mathbf{n}\,\mathbf{m} + \mathbf{m}\,\mathbf{n})$$

$$= \frac{1}{14}\,((3\,e_1 + e_2)(3\,e_1 + 2\,e_2) + (3\,e_1 + 2\,e_2)(3\,e_1 + e_2))$$

$$= \frac{1}{14}\,(9 + 6\,e_{12} + 3\,e_{21} + 2 + 9 + 3\,e_{12} + 6\,e_{21} + 2)$$

$$= \frac{1}{14}\,(22 + 9\,e_{12} + 9\,e_{21})$$

$$= \frac{13}{14} \approx 0.6547 \qquad \Rightarrow \qquad \beta = 21.79°$$

,α elgna noitator eht fo flah eht yltcaxe si β elgna roiretni sihT
:β elgna roiretni eht gnilbuod yllacirtemonogirt nehw nees eb osla nac hcihw

$$\cos(2\,\beta) = 2\cos^2\beta - 1 = 2\cdot\frac{13^2}{14^2} - 1 = \frac{338-196}{196} = \frac{71}{98} = \cos\alpha$$

:noisulcnoC

.noitcelfer fo sexa eht neewteb elgna eht sa gib sa eciwt si noitator fo elgna ehT

!noisulcnoc siht ekil ton od yllaer snaicitamehtam laer tuB
.raelc dna etercnoc oot yaw si tI

,ylthgisnu ,desnednoc ylhgih ,ecracs ,trohs ti ekil snaicitamehtam laeR
.elbisherpmocni neve ro – eguagnal gnisufnoc a ni
.snoitazilausiv fo eerf lytulosba dnA

,noitcelfer ro sexa yb snoitator ebircsed ton yllausu od snaicitamehtam eroferehT
:R srotor otni meht enibmoc yeht tub

$$\mathbf{R} = \mathbf{n}\ \mathbf{m}$$

:eb lliw melborp elpmaxe ruo fo **R** rotor eht suhT

$$\mathbf{R} = \mathbf{n}\ \mathbf{m} = \frac{1}{7}(3\ e_1 + e_2)(3\ e_1 + 2\ e_2) = \frac{1}{7}(9 + 6\ e_{12} + 3\ e_{21} + 2) = \frac{1}{7}(8 + 3\ e_{12})$$

.eno yltcaxe fo htgnel a evah hcihw srotcev edis htiw sesubmohr era srotor oS

.8 retpahc ni "sesubmohr tinu" dellac neeb ylippols dah sesubmohr esehT

dnarepo detator = (submohr tinu) dnarepo (submohr tinu)

(deniart-esuoh ton llits tub) devorpmi erom tib elttil a yllacitamehtaM
:dias eb dluos ti

dnarepo detator = rotor dnarepo rotor

,tcerroc yletelpmoc ton si siht tuB
.detrotsid si tcudorp hciwdnas eht fo srotcaf eht fo redro eht esuaceb

:si tsuj snoitator gnibircsed noitauqe terroc ,tcefrep ehT

$$\mathbf{r}_{rot} = \mathbf{n}\ \mathbf{m}\ \mathbf{r}\ \mathbf{m}\ \mathbf{n}$$

:eb ot sah suht tcudorp hciwdnas rotor a otni noitalsnart ehT

$$\tilde{\mathbf{R}} = \mathbf{m}\ \mathbf{n} \qquad \text{htiw} \qquad \mathbf{r}_{rot} = \mathbf{R}\ \mathbf{r}\ \tilde{\mathbf{R}}$$

melborp elpmaxe ruo ni oS
:si srotcaf sti fo redro degnahcretni htiw rotor desrever eht

$$\tilde{\mathbf{R}} = \mathbf{m}\ \mathbf{n} = \frac{1}{7}(3\ e_1 + 2\ e_2)(3\ e_1 + e_2) = \frac{1}{7}(9 + 3\ e_{12} + 6\ e_{21} + 2) = \frac{1}{7}(8 + 3\ e_{21})$$

noitator nwonk ydaerla eht ot sdael rotor sihT
:noitaluclac ylthgisnu dna ,desnednoc ylhgih ,ecracs ,trohs a htiw

58

$$\mathbf{r_{rot}} = \mathbf{R}\,\mathbf{r}\,\tilde{\mathbf{R}} = \frac{1}{7}(8 + 3\,e_{12})(2\,e_1 + e_3)\frac{1}{7}(8 + 3\,e_{21})$$

$$= \frac{1}{49}(13\,e_1 + 11\,e_3)(8 + 3\,e_{21})$$

$$= \frac{1}{49}(71\,e_1 + 94\,e_3) \approx 1.45\,e_1 + 1.92\,e_3$$

elbisneherpmocni erom hcum osla tub ,retrohs hcum si noitaluclac sihT
.retpahc tsal eht ni noitaluclac eht ot derapmoc

ydorap noinretauq sihT
.noitator a fo erutcurts cirtemoeg laer eht dna gninaem laer eht sedih ylevitceffe yrev
wolla lliw snoinretauq eurt snotlimaH eroferreht dnA
.snoitator fo snoitatupmoc evitceffe yrev

.deneppah tahw dnatsrednu lliw ydobon tuB
eroc lautpecnoc eurt eht tuoba egdelwonk dilos dna raelc a teg lliw ydoboN
.snoitator fo erutcurts gniylrednu eurt eht dna

:(tniopweiv lacirotsih) noisulcnoC

,suineg a saw notlimaH
snoitator gnibircsed ni dedeeccus eh esuaceb
.snoitcelfer gnisu tuohtiw dna gnidnatsrednu tuohtiw

.nward eb tsum noisulcnoc rehtona ,emit emas eht tA

:(tniopweiv lacitcadid) noisulcnoC

,loof suoinegni na saw notlimaH
dedeeccus ton sah eh esuaceb
.snoitcelfer gnisu yb snoitator gnibircsed ni dna gnidnatsrednu ni

15 Outlook: Algebra of Tetrapods

Of course Hamilton was able to describe rotations in three-dimensional space.
But we are not able to model three-dimensional situations
by using the algebra of intelligent starfishes.
Here everything will take place in one single plane.

To model three-dimensional situations without negative expressions mathematically,
four unit vectors e_1, e_2, e_3, and e_4 are required. These four unit vectors
then point in form of a tetrapod into the three directions of space.

Thus an algebra of intelligent tetrapods is required.

An intelligent, but unfortunately rather lethargic tetrapod from Darmstadt
is shown on the picture at the bottom of this page.
The four podgy ends of the tetrapod form a tetrahedron.

(c) Copyright by the author, own photograph taken in front of the Institute of Hydraulic
Engineering of TU Darmstadt at January 13th, 2012.

nettirw eb nac srotcev tinu ruof eht suhT

arbeglA iluaP ro arbeglA cirtemoeG lanoisnemid-eerht fo egaugnal eht ni

sa

$$e_1 = \sigma_z$$

$$e_2 = \frac{1}{3}\sqrt{8}\,\sigma_x - \frac{1}{3}\sigma_z$$

$$e_3 = -\frac{1}{3}\sqrt{2}\,\sigma_x + \frac{1}{3}\sqrt{6}\,\sigma_y - \frac{1}{3}\sigma_z$$

$$e_4 = -\frac{1}{3}\sqrt{2}\,\sigma_x - \frac{1}{3}\sqrt{6}\,\sigma_y - \frac{1}{3}\sigma_z$$

:si neht mus llun eht dnA

$$e_1 + e_2 + e_3 + e_4 = 0$$

,snoitauqe latnemadnuf ehT

,iluaP fo snoitaler noitatummoc eht ot ralimis era hcihw

yfsitas neht

$$e_1 e_2 + e_2 e_1 + \frac{2}{3} = 0 \qquad e_2 e_3 + e_3 e_2 + \frac{2}{3} = 0$$

$$e_1 e_3 + e_3 e_1 + \frac{2}{3} = 0 \qquad e_2 e_4 + e_4 e_2 + \frac{2}{3} = 0$$

$$e_1 e_4 + e_4 e_1 + \frac{2}{3} = 0 \qquad e_3 e_4 + e_4 e_3 + \frac{2}{3} = 0$$

,snoitauqe eseht no desaB

,detcurtsnoc eb nac erutcurts lacitamehtam etelpmoc a

.sngis sunim fo eerf yletelpmoc si hcihw dna seititnauq evitisop sah ylno hcihw

,yrots rehtona si siht tuB

.liated retaerg ni koob rehtona ni dlot eb lliw hcihw

16 Literature

[1] Albrecht Beutelspacher: „In Mathe war ich immer schlecht ...", 5th edition, Vieweg + Teubner / GWV Fachverlage, Wiesbaden 2009.

[2] David Hestenes: Oersted Medal Lecture 2002: Reforming the mathematical Language of Physics. In: American Journal of Physics, Vol. 71, No. 2 (2003), pp. 104 – 121.

[3] Stephen Gull, Anthony Lasenby, Chris Doran: Imaginary Numbers Are Not Real – The Geometric Algebra of Spacetime. In: Foundations of Physics, Vol. 23, No. 9 (1993), pp. 1175 – 1201.

[4] Martin Erik Horn: Grass, Mann! Das Clifford-Kinder-Rechenbuch. In: Volkhard Nordmeier, Arne Oberländer (Eds.): Didaktik der Physik, Beiträge zur Frühjahrstagung in Düsseldorf, Tagungs-CD des Fachverbands Didaktik der Physik der Deutschen Physikalischen Gesellschaft, LOB – Lehmanns Media, Berlin 2004, ISBN 3-86541-066-9.

[5] Hermann Graßmann: Die Wissenschaft der extensiven Größe oder die Ausdehnungslehre, eine neue mathematische Disciplin. Erster Theil, die lineale Ausdehnungslehre enthaltend. Verlag von Otto Wigand, Leipzig 1844.

[6] Hermann Graßmann: Die Ausdehnungslehre. Vollständig und in strenger Form bearbeitet. Verlag von Th. Chr. Fr. Enslin, Berlin 1862.

[7] George Sarton: Grassmann – 1844. In: Isis, Vol. 35, No. 4 (Autumn 1944), pp. 326 – 330.

[8] David Hestenes: New Foundations for Classical Mechanics. 2nd edition, Kluwer Academic Publishers, New York, Boston, Dordrecht 2002.

[9] John Snygg: Clifford Algebra. A Computational Tool for Physicists. Oxford University Press, New York, Oxford 1997.

[10] Chris Doran, Anthony Lasenby: Geometric Algebra for Physicists. Cambridge University Press, Cambridge 2003.

[11] Leo Dorst, Daniel Fontijne, Stephen Mann: Geometric Algebra for Computer Science. An Object-oriented Approach to Geometry. Morgan Kaufmann Publishers / Elsevier, Amsterdam, Boston, Heidelberg, 2007.

[12] John Vince: Geometric Algebra for Computer Graphics. Springer-Verlag, London 2008.

[13] Dietmar Hildenbrand: Introduction to Geometric Algebra Computing. CRC Press / Taylor & Francis Group, Boca Raton, London, New York 2019.

[14] Martin Erik Horn: Pauli-Algebra und S_3-Permutationsalgebra – Eine algebraische und geometrische Einführung, www.bookboon.com / Ventus Publishing ApS, London 2012.

[15] Martin Erik Horn: Eine andere Geometrische Algebra. In: PhyDidB – Didaktik der Physik, Beiträge zur DPG-Frühjahrstagung des Fachverbands Didaktik der Physik in Mainz 2012, Beitrag 02.7.

[16] Martin Erik Horn: Die Raumzeit-Algebra der (3 x 3)-Matrizen.

,nenreL sednehcsroF – gninraeL desab-yriuqnI :(.dE) tlohnreB ahcsaS :nI
,322 – 320 .pp ,33 dnaB ,revonnaH ni PCDG red gnugatserhaJ ruz egärtieB
,(NPI) kitamehtaM dnu netfahcsnessiwrutaN red kigogadäP eid rüf tutitsnI
.2013 leiK
ni netkudorP nereßuä dnu nerenni nehcsiwz gnuheizeB ruZ :nroH kirE nitraM [17]
nitraM , kcinpäK mlehdeirF ,htarfeerG trebliG :nI .arbeglA nehcsirtemoeG red
,2013 thcirretnukitamehtaM muz egärtieB – UMzB :(.sdE) nietS
,1 dnaB ,kitamehtaM red kitkadiD rüf tfahcslleseG red dnabsgnugaT
.2013 retsnüM ,galreV-MTW ,483 – 480 .pp

:dnemmocer ot oslA

.noisnemiD etreiv eid dnu sarogahtyP :nroH kirE nitraM [18]
red gnureniemegllareV ehcsirtemoeg eid rebü kcilbrebÜ niE
.sevlaM ed uaG ed dnu sarogahtyP nov neppurgztaS
.978-3-7562-0388-8 :NBSI ,2022 tdetsredroN ,dnameD no skooB – DoB

PYTHAGORAS
UND DIE VIERTE
DIMENSION

sloohcs ta dlot si sarogahtyP fo meroehT ehT
.esac laiceps ralacs thguartsid dna , ylenol desserped ,das a sa
:yllof namuh a si tcelgen egnarts siht rof snosaer ynam eht fo enO
,latnemadnuf dna ,ysae ,elpmis taht si sarogahtyP fo meroehT lairotcev ehT
.ti ees ot deliaf evah ew won litnu taht

:si sarogahtyP fo meroehT lairotcev eurt ehT

a + b = c

.elgnairt elbissop yreve rof ,erehwyreve ,emit eht lla dilav si noitauqe lairotcev sihT
mrof latnemele ,lanigiro ,laitnesse ,tnerehni eht si ti dnA
,dilcuE dna sarogahtyP fo smeroehT eht fo
.smeroeht eseht fo noitadnuof dilos etinifni eht dna ecnesse lautca ,eurt eht gnitats
:elbat siht fo snoitauqe eht teg ot ti erauqs ot evah ylno eW

selgnairT thgiR		selgnairT lareneG	
$c^2 = a^2 + b^2$		$c^2 = a^2 + b^2 + 2\,a \bullet b$	
$p\,c = a^2$	$q\,c = b^2$	$p\,c = a^2 + a \bullet b$	$q\,c = b^2 + a \bullet b$
$p\,q = h^2$	$a\,b = c\,h$	$p\,q = h^2 + a \bullet b$	$a\,b = c\,h + a \bullet b$

:no seog yrots eht dnA

(namreG ni nettirw) "noisnemiD etreiv eid dnu sarogahtyP" koob eht nI
dilcuE dna sarogahtyP fo smeroeht eht fo snoitazilareneg lanoisnemid-rehgih eht
.snorohcatnep lareneg dna snordehartet lareneg rof nevig era